Praise for *The Green-Collar Economy*

"This book illustrates the link between the struggle to restore the environment and the need to revive the U.S. economy. Van Jones demonstrates conclusively that the best solutions for the survivability of our planet are also the best solutions for everyday Americans."

—Al Gore, former vice president and cofounder
and chairman of Generation Investment
Management

"As the Earth warms and the oceans rise, the civil and human rights agenda must expand. In the wake of Katrina and Rita, we must focus on equal protection from environmental disasters. And as solar and wind industries take off, our jobs agenda must embrace these new opportunities. No one has worked harder to level the playing field in the rapidly growing green economy than Van Jones."

—Ben Jealous, president, NAACP

"The baton is passed to climate advocate Van Jones who clearly sees that our future must be green and must include everyone. His powerful new book *The Green-Collar Economy* shows us how to accomplish it."

—Laurie David, global warming activist,
stopglobalwarming.org

"Van Jones reminds us that the worst of times can also be the best of times; that a nation with an abundance of resources it's wasting—beginning with its youth—has an enormous opportunity to stop foolishly bankrupting itself by chasing resources it is running out of, such as oil. We also have lots of ingenuity, engineering talent, wind, and sun—put them together and we have a powerful tomorrow."

—Carl Pope, executive director, Sierra Club

"With *The Green-Collar Economy*, Van Jones accomplishes the super heroic feat of linking together the solutions for poverty, the energy crisis, and global warming. That's 'silver rights' movement leadership at its best. And Van is the only environmentalist I know who acknowledges the central role of low-wealth communities in creating the green world of the future. Van is a visionary of our times, and one of my personal heroes. Every relevant 21st century leader needs to read Van's book."

—John Hope Bryant, founder & CEO,
Operation Hope

"Van's words echo the sentiments of many indigenous communities, who have endured the effects of coal strip mining, uranium mining, and mega dams. We too embrace a green path for the future—indeed this green path is a part of Ojibwe prophecies. There is an absolute mandate in the wealthiest society in the world to create an economy not based on conquest but on survival, and on dignity between peoples and the natural world. *The Green-Collar Economy* outlines industrial society's path towards a just future."

—Winona LaDuke, environmental and
Native American rights activist

"Around the world, people of African descent are creating exciting new environmental movements: from Kenya's Wangari Maathai to the South Bronx's Majora Carter. To that list, we now can add a new name: Van Jones. In *The Green-Collar Economy*, he shows how 'green' can be good for people of ALL colors."

—Kerry Washington, actor

"It's rare that someone with such a gift for speaking is able to convey the energy and excitement of his message equally well in writing. With *The Green-Collar Economy*, Van Jones surpasses all expectations. The country seriously needs his take on the environment and the economy."

—Gavin Newsom, mayor of San Francisco

"In *The Green-Collar Economy*, Van Jones has penned a working folks' manifesto for the solar age. When green solutions finally catch on among everyday people, Van and this book will deserve the lion's share of the credit."

—Rev. Lennox Yearwood, Hip Hop Caucus

"*The Green-Collar Economy* is a both a rallying call and a road map for how we can save the planet, reduce our dependency on budget-busting fossil fuels, and bring millions of new jobs to America. Van Jones shows how climate solutions can turbocharge the ailing U.S. economy. So what are we waiting for?"

—Fred Krupp, president of the Environmental Defense Fund and *New York Times* best-selling coauthor of *Earth: The Sequel*

"Brother Van Jones is a visionary who spells out real solutions in black and white—and, of course, green. Van's vision of a thriving, green economy doesn't have throw-away things or throw-away people. It's the kind of environmentalism everyone can get behind."

—Mario Van Peebles, actor and producer, *Mario's Green House*

"I can think of no one more pertinent to our future than Van Jones and his seamless integration of race, poverty, the environment, and human rights."

—Paul Hawken, author of *Blessed Unrest*

"Out of those seen as today's leaders, I know Van Jones to be one of the most important visionaries, strategists, and voices of our times."

—Julia Butterfly Hill, eco-heroine and author of
The Legacy of Luna

"Every once in a while, in the course of history, someone comes along—often miraculously and just in time—who completely changes the game. Van Jones is that person, and his breathtaking new book, *The Green-Collar Economy* will do exactly that. Van has written an absolutely stunning, astounding, and life-changing book."

—Lynne Twist, author of *Soul of Money*

THE
GREEN-
COLLAR
ECONOMY

Contents

green-collar job
\\'grēn-'kä-lər 'jäb\\ noun

: blue-collar employment that has been upgraded to better respect the environment
: family-supporting, career-track, vocational, or trade-level employment in environmentally-friendly fields
: examples: electricians who install solar panels; plumbers who install solar water heaters; farmers engaged in organic agriculture and some bio-fuel production; and construction workers who build energy-efficient green buildings, wind power farms, solar farms, and wave energy farms

Foreword

ROBERT F. KENNEDY JR.

L AST NOVEMBER, LORD (David) Puttnam debated before Parliament an important bill to tackle global warming. Addressing industry and government warnings that we must proceed slowly to avoid economic ruin, Lord Puttnam recalled that precisely two hundred years ago Parliament heard identical caveats during the debate over abolition of the slave trade. At that time slave commerce represented one-fourth of Britain's gross domestic product (GDP) and provided its primary source of cheap, abundant energy. Vested interests warned that financial apocalypse would succeed its prohibition.

That debate lasted roughly a year, and Parliament, in the end, made the moral choice, abolishing the trade outright. Instead of collapsing, as slavery's proponents had predicted, Britain's economy accelerated. Slavery's abolition exposed the debilitating inefficiencies associated with zero-cost labor; slavery had been a ball and chain not only for the slaves, but also for the British economy, hobbling productivity and stifling growth. Creativity and productivity surged. Entrepreneurs seeking new sources of energy launched the industrial revolution and inaugurated an era of the greatest wealth production in human history.

Today, we don't need to abolish carbon as an energy source in order to see its inefficiencies starkly or understand that the addiction to it is the principal drag on American capitalism. The evidence is before our eyes. The practice of borrowing a billion dollars each day to buy foreign oil has caused the American dollar to implode. More than a trillion dollars in annual subsidies to coal and oil producers has beggared a nation that four decades ago owned half the globe's wealth. Carbon dependence has eroded our economic power, destroyed our moral authority, diminished our international influence and prestige, endangered our national security, and damaged our health and landscapes. It is subverting everything we value.

We know that nations that "decarbon" their economies reap immediate rewards. Sweden announced in 2006 the phaseout of all fossil fuels (and nuclear energy) by 2020. In 1991 the Swedes enacted a carbon tax (now up to $150 a ton), closed two nuclear reactors, and still dropped greenhouse-gas emissions to five tons per person, compared to the U.S. rate of twenty tons. Thousands of entrepreneurs rushed to develop new ways of generating energy from sun, wind, and tides and from wood chips, agricultural waste, and garbage. Growth rates climbed to upwards of three times those of the United States. The heavily taxed Swedish economy is now the world's eighth richest by GDP.

Iceland was 80 percent dependent on imported coal and oil in the 1970s, its economy among the poorest in Europe. Today, Iceland is 100 percent energy-independent, with 90 percent of the nation's homes heated by geothermal and its remaining electrical needs met by hydro. The International Monetary Fund now ranks Iceland the fourth most affluent nation on Earth. Geothermal and hydro produce so much cheap power that Iceland has become one of the world's top energy exporters. (Iceland exports its surplus energy in the form of smelted aluminum.) The country, which previously had to beg for corporate investment, now has companies lined up to relocate there to take advantage of its low-cost clean energy.

Brazil, which decarboned its energy over the past decade, is now experiencing the most sustained economic boom in its history. Costa Rica, which is phasing out carbon, is Central America's wealthiest economy. It should come as no surprise that California, America's most energy-efficient state, also possesses its strongest economy.

The United States has far greater domestic energy resources than Iceland or Sweden. We sit atop the second-largest fund of geothermal resources in the world. The American Midwest is the Saudi Arabia of wind; indeed, North Dakota, Kansas, and Texas alone produce enough harnessable wind to meet all of the nation's electricity demand. As for solar, according to a study in *Scientific American,* photovoltaic and solar-thermal installations across just 19 percent of the most barren desert land in the Southwest could supply nearly all of our nation's electricity needs without any rooftop installation, even assuming every American owned a plug-in hybrid car. This is, incidentally, a much smaller footprint than would be required by the equivalent power from coal.

In America, several obstacles impede the kind of entrepreneurial revolutions that brought prosperity to Sweden and Iceland. First, that trillion dollars in annual coal and oil subsidies gives the carbon industry a decisive market advantage and creates a formidable barrier to renewables. Second, an overstressed and inefficient national electrical grid can't accommodate new kinds of power. Third, a byzantine array of local rules impedes access by innovators to national markets. And fourth, state and federal governments have failed to develop efficiency standards and long-promised market incentives for green buildings and machines.

There are four things the new president should immediately do to hasten the approaching boom in energy innovation. A carbon cap-and-trade system designed to put downward pressure on carbon emissions is quite simply a no-brainer. Already endorsed by Senators John McCain and Barack Obama, such a system would measure national carbon emissions and create a market to auction emissions credits. The supply of credits is then reduced each year to meet predetermined carbon-reduction targets. As supply tightens, credit value increases, providing rich monetary rewards for innovators who reduce carbon. Since it is precisely targeted, cap-and-trade is more effective than a carbon tax. It is also more palatable to politicians, who despise taxes and love markets. Industry likes the system's clear goals. This market-based approach has a proven track record.

The next president must push to revamp the nation's antiquated high-voltage power-transmission system, so that it can deliver solar, wind, geothermal, and other renewable energy across the country. Right now, a Texas wind-farm manager who wants to get his electrons to market faces two huge impediments. First, our regional power grids are overstressed and misaligned. The biggest renewable-energy opportunities—for instance, Southwest solar and Midwest wind—are outside the grids' reach. Furthermore, traveling via alternating-current (AC) lines, too much of that wind farmer's energy would dissipate before it crossed the country. The nation urgently needs more investment in its backbone transmission grid, including new direct-current (DC) power lines for efficient long-haul transmission. Even more important, we need to build in "smart" features, including storage points and computerized management overlays, allowing the new grid to intelligently deploy the energy along the way. This backbone would operate at the speed of light and incorporate sophisticated new battery and storage technologies to store solar energy for use at night and to deploy wind energy during the doldrums. Construction of this new grid will create a marketplace where utilities, established businesses, and entrepreneurs can sell energy and efficiency.

The other obstacle is the web of arcane and conflicting state rules that currently restrict access to the grid. The federal government needs to work with state authorities to open up the grid, allowing clean-energy innovators to fairly compete for investment, space, and customers. We need open markets where hundreds of local and national power producers can scramble to deliver economic and environmental solutions at the lowest possible price. The energy sector, in other words, needs an initiative analogous to the 1996 Telecommunications Act, which required open access to all the nation's telephone lines. Marketplace competition among national and local phone companies instantly precipitated the historic explosion in telecom activity.

Construction of efficient and open-transmission marketplaces and a green-power-plant infrastructure would require about a trillion dollars over the next fifteen years. For roughly a third of the projected cost of the Iraq war we could wean the country from carbon. And the good news is that the government doesn't actually have to pay for all of this. If the president works with governors to lift constraints and encourage investment, utilities and private entrepreneurs will quickly step in to revitalize the grid and recover their investment through royalties collected for transporting green electrons.

One investor anxious to fill this breach is Stephan Dolezalek, a managing director of VantagePoint Venture Partners, one of the world's largest green-tech venture-capital firms. Dolezalek scoffs at claims that a carbon-free economy is still decades away. "With the right market drivers and an open-access marketplace, we can completely decarbon our electric system within years," says Dolezalek. He analogizes the grid initiative to the federal Arpanet high-speed

Internet backbone that accelerated the PC revolution and the information-technology boom in the 1990s. "In 1987, there were less than 500 networks," he recalls. "By 1995, there were 50,000. By 1996, there were 150,000. The energy sector has the potential to evolve forward more quickly than most people can grasp today. We're going to see those same quick responses in the renewable-energy sector. As soon as the national marketplace is up, the curves will go vertical."

Energy expert and former CIA director R. James Woolsey predicts: "With rational market incentives and a smart backbone, you'll see capital and entrepreneurs flooding this field with lightning speed." Ten percent of venture-capital dollars is already deployed in the clean-tech sector, and the world's biggest companies are crowding the space with capital and scrambling for position. Says Dolezalek, "The Internet boom caused information flow to increase exponentially, but the price per bit dropped to almost zero. The same thing can happen with energy." Dolezalek reminds us that energy is hitting the Earth for free. We just need to erect the infrastructure to harvest and deliver it to the consumers. Solar and wind plants are far quicker to deploy than conventional power plants because of their simple design and lower environmental-impact concerns. The plants have modest maintenance and operation costs. There are no mining, refining, or transportation costs or the catastrophically expensive environmental and military consequences associated with carbon.

"We have the ability to make clean energy both abundant and cheap," says Dolezalek. Accessible markets will give every American the opportunity to become an energy entrepreneur. Homes and businesses will become power plants as individuals cash in by installing solar panels and wind turbines on their buildings and selling the stored energy in their plug-in hybrids back to the grid at peak hours. "As capital and entrepreneurs rush into this space, the pace of change will accelerate exponentially. As energy production goes up, you could see the price per unit drop to practically nothing."

The president's final priority must be to connect a much smarter power grid to vastly more efficient buildings and machines. We have barely scratched the surface here. Washington is a decade behind its obligation, first set by Ronald Reagan, to set cost-minimizing efficiency standards for all major appliances. With the conspicuous exception of Arnold Schwarzenegger's California, the states aren't doing much better. And Congress keeps setting ludicrously tight expiration dates for its energy-efficiency tax credits, frustrating both planning and investment. The new president must take all of this in hand at once.

We need to create open national markets where individuals who devise new ways to produce or conserve power can quickly profit from their innovations. Open, efficient markets will unleash America's entrepreneurial energies to solve our most urgent national problems—global warming, national security, our staggering debt, and a stagnant economy. Everyone will profit from the green gold rush. By kicking its carbon addiction, America will increase its national wealth and generate millions of jobs that can't be outsourced. We will create a decentralized and highly distributable grid that is far more resilient and safe for our country; a terrorist might knock out a power plant, but never a million homes. We will cut annual trade and budget deficits by hundreds of billions and improve public health and farm production. And for the first time in half a century we will live free from Middle Eastern wars and entanglements with petty tyrants who despise democracy and are hated by their own people.

The Green-Collar Economy is a critical step forward into this brave new world. Van Jones articulates the urgency and importance of the task and the opportunity before us. Let the revolution begin.

Reality Check

F IRST THE BAD news: decades of shortsighted economic and environmental policies have torn the floor out from under the American people. The poorest among us are most at risk, but the future for everyone looks grim.

The world economy and the polar ice caps are both melting down. The global recession is bowling over investments and battering down job prospects. The impacts of global warming are imperiling our forests, our farmland, and our future. There is no place to hide—from either the economic crisis or the ecological threat. Here and around the world, humanity faces an unprecedented double danger on a massive scale.

The solution is challenging but straightforward: we must create a new, "green-collar economy"—one that will create good, productive jobs while restoring the health of our planet's living systems. By taking bold action, we can turn these two terrifying breakdowns into a single, comprehensive breakthrough.

Such a transformation is possible, on both a national and a global scale. That's the good news. A big reason for optimism is that on

November 4, 2008, millions of Americans—seeing the economic floor collapsing beneath their feet—marched into voting booths and decided to tear the political ceiling off too.

So, yes, it is true that the floor is gone. But thanks to the election of Barack Obama, so is the ceiling. Today we can fall—or we can fly. It is up to us; that decision is in our hands. The green economic remedies described in this book can help us begin to soar. But before we start designing our new flight plan, we should take a moment to understand why we started falling in the first place.

Obama's presidency opens the door to a holistic set of solutions. But make no mistake: the problems did not originate with George W. Bush (though his misdeeds and incompetence added incalculable pain and damage). The sad truth is that for decades both political parties have promoted economic programs based on three fundamental fallacies—and both parties share in the blame for our present predicament. Democrats and Republicans together assured the American public that we could grow our economy based on: (1) consumption rather than production, (2) credit rather than thrift, and (3) ecological destruction rather than ecological restoration. The present crisis has exposed all three of these notions as dangerous shams.

The first fallacy was the idea that we could continue to be the world's biggest and most important economic power not by being the world's biggest producer, but by being the world's biggest consumer. For years, both parties assured the public that the continual shipping of U.S. manufacturing jobs overseas by multinational corporations was really no big deal. In fact, our elected leaders continually pushed through trade agreements that would accelerate the outsourcing of American jobs to places with pitifully low wages and environmental standards. Occasionally ordinary Americans began to fret and grumble. But a bipartisan chorus told us, in effect: "Don't worry! Let all the other countries have your old blue-collar jobs. You will get something *better!"* Unfettered free trade would give our country a "new economy"—one based on providing information and other services—and everything would work out fine.

Sure enough, although production of real goods began to shift overseas, the United States did maintain a central role in the world economy. But our role was as a nation of consumers rather than a nation of producers. The countries of the world still needed us—not to build things for them, but to buy things *from* them. Once again, the people who voiced doubts were shouted down. So most Americans went along with the plan. We dutifully traded in our sacred symbols. We stopped putting our faith in the skilled American worker holding a tool; instead, we enshrined the online shopper clutching a mouse. In the new American mythos, the temple of economic progress was no longer the American factory, but the American shopping mall. And we hoped that everything would turn out as our elected officials promised.

The new economy did not work as advertised. Millions of Americans today find themselves not in a consumer's paradise, but in a worker's hell. They are trying to eke out a living by changing bedpans, making lattes, or stocking shelves in big-box stores with products made by workers overseas. We have a situation in which the trade is "free"—but the people are not. The simple truth is that no advanced industrial country can go on for long consuming more than it produces without its economy eventually running aground—along with many of its people's hopes and dreams.

We did not recognize the danger of the first fallacy, because we were blinded by the appeal of the second: the notion that we could go forward indefinitely by relying on excessive debt and credit rather than on smart savings and thrift (the way our grandparents did). Easy credit, often tied to the real-estate bubble, let millions of us consume and consume, often far beyond our means, for years. Low-interest credit-card offers came flying at us, unbidden, from our mailboxes.

So did invitations to refinance our homes. The temptation to indulge was overwhelming. Homeowners began to hock their houses just to buy flat-screen TVs (perhaps to cover the holes in their lives). The concept of saving for years to make a down payment on a modest home began to look silly in the era of "EZ" home loans for everyone. The very idea that people should save up their money to make a major purchase began to look old-fashioned. Why heed grandma's "save today, buy tomorrow" counsel, when our magical credit cards would let us have everything now, now, now? Average U.S. savings eventually plunged below zero, as America indulged in an orgy of consumption. The world cheered and happily shipped us more junk to buy. And since our nation's financial regulators were asleep at the switch, some thought the party would never end.

But it did. The bubble burst. The bills came due. And now, for millions of people, those credit-card offers have been replaced by demands for payment, overdraft warnings, and—too often—eviction notices. No nation can go on forever relying on overseas credit rather than domestic creativity, on borrowing rather than building.

The final fallacy was that we could drive economic growth forever through ecological destruction rather than ecological restoration. For decades, to create strip malls and housing sprawl, we have been carving up precious wilderness and paving over needed farmland. We have been wasting priceless resources like freshwater, while fouling the skies with pollution. Our factory farms maintain themselves using massive amounts of antibiotics and pesticides, all of which wind up inside human beings. For too long we have been pillaging our living ecosystems to make dead products, most of which we quickly dump or incinerate. The whole process produces dirtier air and water, which endanger the health of our children. Epidemic rates of asthma and cancer have become the norm. At some point, these imbalances threaten to catch up with us—in the form of catastrophic public-health disasters and resource shortages.

Nothing, however, is more destructive than the way we have chosen to power our society—by mining, drilling for, and burning fossil fuels. We have failed to invest in stable, domestic, clean energy sources. As a result, we have exposed ourselves to volatile energy prices—and even more volatile weather patterns. In recent years, extreme weather events—from hurricanes in the Gulf Coast area, to floods in Arkansas and Iowa, to wildfires in California—have given us visual evidence that Al Gore is right: greenhouse-gas pollution from burning fossil fuels may already be disrupting our climate. Some ecological bills, apparently, have already come due. Our scientists tell us that even bigger and scarier ones are still on their way.

Thus we find ourselves at the end of an era of American capitalism—one of unregulated growth. It was a spectacular period, but in the end it largely failed to deliver on its promises. The only way to return to the path of sanity and progress is simple: we need to turn each of these three fallacies on its head. We need to go back to producing things here again. We need to go back to relying on smart savings and thrift—including conserving our natural resources. And we need to start honoring the Earth rather than just plundering it.

We can do all three. The centerpiece of such an effort would be a crash program for energy independence based on clean energy and energy efficiency. By producing renewable energy and other green products here, while better conserving our monetary and natural resources, we can create secure pathways to more work, more wealth, and better health for millions of Americans.

Some people may pooh-pooh this idea, seeing it as too idealistic or too ambitious. Some may imagine that we can get our economy back on track without seriously addressing our overreliance on outdated energy sources or the broader ecological peril in which we find ourselves. These people fail to grasp the need for a profoundly new direction. The fact is, a return to the good old days of "go-go" hypercapitalism is probably not a realistic option for the United States. But even if we could go back to precrash "business as usual," we would be crazy to do so—especially when it comes to our energy policy.

In the following pages, we explore the economic dangers of our continued reliance on oil and other nonrenewable fuel sources. I point out viable, clean-energy alternatives, expose some false choices, and explain the concept of green-collar jobs as the first step toward laying out a vision for a better future.

THE PATH TO ECONOMIC RUIN IS SLICK WITH OIL

It is worth spelling out the dangers we would face, were we to somehow revive the old status quo. Because our society has remained dependent on oil in every aspect of our lives, petroleum prices remain the Achilles' heel of the entire economy. This weakness continues to have the potential to send the entire country into a tailspin.

This tailspin has a name, one that sends shivers of horror down the spine of every economist: stagflation—stagnant economic growth occurring simultaneously with runaway inflation. This horrible economic malady, sparked by high gas prices, made the 1970s dismal. In the summer of 2008, fuel prices began climbing again—and the specter of stagflation haunted the nation. Rather than stagflation, we got a global economic collapse that sent oil prices tumbling down again. But sooner or later, global demand for oil will outstrip supply; oil price rises are thus inevitable. And so is the risk of stagflation.

Stagflation is perhaps the worst possible outcome in a market economy. It is rare, yet it is almost always fueled by a sharp rise in energy prices. To put it simply, in a stagflation scenario the prices keep going up—and the number of jobs keeps going down.

The reason is straightforward. It takes energy to make anything and everything. So when energy costs go up, all prices tend to go up. At the same time, those very same steep energy prices eat into consumer confidence. They depress nonessential spending and discourage hiring. So consumers stop buying, employers hold back on making job offers, and tourists travel much less. As a result, the economy starts to stall—with all the attendant job loss and pain. Yet prices throughout the economy, driven by rising fuel costs, keep going up just the same. The result is that society finds itself stretched on the rack, with soaring costs and plunging jobs pulling the body of the nation in opposite directions.

It is nearly impossible to grow the economy and add jobs when energy prices are going through the roof. And there is no easy way out of this inflationary cycle. Of course, over time, sky-high energy prices will force both individuals and businesses to consume less energy and seek alternatives; a new equilibrium point will be found. Yet by the time this kind of shift occurs naturally, an oil-dependent economy like ours could be dead as a doornail.

The solution for the economy is simple: head off stagflation down the line by deliberately cutting demand for energy and intelligently increasing its supply today. Those two steps will bring overall energy supply and demand back into balance, stabilizing energy costs and eventually lowering them.

But all of that is a lot easier said than done. Our economy is powered almost exclusively by fossil fuels, a nonrenewable resource. That means the supplies are limited—by definition. There is only so much oil, natural gas, and coal in the world.[1] The more we use the stuff, the less of it we have—and the more it will cost us over time. The laws of supply and demand tend to make dwindling resources more and more expensive over time.

Unfortunately, our entire economy was designed to function in a world where fossil fuels are forever abundant and forever cheap.

Today, as those fuels—especially oil—become increasingly scarce, prices will rise to reflect that reality.

And they will keep rising. The main reason is that the oil supply can no longer hope to keep pace with demand. On the one hand, global demand is set to skyrocket, especially from growing economies such as those of India and China. As both countries blossom into full-blown, economic superpowers, energy demand from the Asian continent will only increase.

However, on the supply side, oil companies are not finding more oil fields.[2] Some experts even fear that global oil production has already peaked and that supplies are headed for a permanent worldwide decline—even as demand goes up. Combine that fact with a weakening dollar, and you have an irreversible spike in U.S. oil and gas prices. In other words, the days of cheap oil and gas are over—forever.[3] So if we want to lower fuel prices by increasing the supply of energy, we must find alternatives to oil—and eventually to coal and natural gas as well.

Of course there are plenty of proposals to keep us addicted to fossil fuels. Some of these ideas are downright scary. Others are bizarre. To keep us hooked on polycarbons, companies are now proposing that we make fuels out of tar sands and oil shale.[4] Or drill off our shores and in our national parks.[5] Or liquefy coal to run our cars.[6] Reputable experts doubt whether these dubious efforts will make more than a small dent in the energy problem, but there seems to be no end to the efforts to find "alternatives" that are actually just a repackaging of the same fossil fuels—with the same problems. What's worse, many of these so-called alternatives come from the polycarbons that are the hardest to extract, costliest to produce, and nastiest to burn. Yet companies are still willing to go out there, dig them up, and try to bring them to market—at the right price.

Further exacerbating the problem, the world's scientists are sounding the alarm that we cannot continue to burn fossil fuels at anywhere near the present rate—let alone introduce even dirtier fuels. Greenhouse-gas levels may have already passed key tipping points, threatening to overheat the atmosphere and unleash climate chaos. The experts posit that we must cut back on burning fossil fuels altogether.[7] That's right. Not only are fuel costs prohibitive and leading us down a path to economic ruin; any attempt to offset those oil prices by burning dirtier fossil fuels would essentially cook the Earth.

And so we find ourselves stuck with a dual crisis and a potentially major dilemma. Should we use even dirtier fossil fuels to rev up the economy and in turn bake the planet? Or should we stop using oil and coal tomorrow and wreck the economy?

Whom do we love and care about more? Our children—and their immediate need for a viable economy? Or our grandchildren—and their long-term need for a viable planet? Go ahead—choose one.

Fortunately, this dilemma is a false choice. It is true that we cannot drill and burn our way out of our present economic and energy problems. We can, however, invent and invest our way out. Choosing to do so on a massive scale would have the practical benefit of cutting energy prices enough—and generating enough work— to pull the U.S. economy out of its present death spiral. But the true benefits would be much greater than that.

A serious shift in our energy strategy would open a new chapter in the story of human civilization. Right now, we are still scurrying about on our planet's surface, eking out our living as part of a vulture society—living off the dead. Out of the Earth we suck the liquefied remains of dead organisms. We burn our ancestors' remains in our engines, without ceremony. Then we go back to the Earth, like vampires, to suck out even more oil. Our coal-fired power plants munch daily on the black bones of the ancients—and belch out death. Today, the climate itself threatens to bring everything full circle: if we keep pulling death from the ground, we will reap death from the skies.

NEW CENTURY, NEW POWER: TAPPING SUN, MOON, AND EARTH'S INNER FIRE

There is a wiser and more civilized alternative. Rather than continuing to base our economy on a finite supply of dead things, we can base it on sources that are practically infinite and eternal: the sun, the moon, and the Earth's inner fire.

Solar energy is as reliable as the sunrise; through solar-thermal and photovoltaics, we can harness the sun's majesty to make abundant, clean energy. Enough solar energy falls on the Earth's surface in one hour to power all of human civilization for a year.[8] The warmth from that same sun creates the weather patterns that also drive the winds; modern wind turbines can turn a gentle breeze into raw power. The interaction between the Earth and the moon creates the tides of the ocean; turbines in the sea can convert their constant rhythm to usable energy, just as wind turbines pull power from the air. And below the Earth's surface, our planet is alive with heat and power. The same drilling technology that once helped us reach pools of dead oil could someday help us tap the living furnace beneath us.

The possibility of an economic recovery based on clean energy (to increase supply) and on wasting less energy (to cut demand) is not a daydream. There is already a huge green economy developing. It is growing despite inadequate and inconsistent support from a public sector that is still easily cowed by the big polluters. In 2006, renewable energy and energy-efficiency technologies generated 8.5 million new jobs, nearly $970 billion in revenue and more than $100 billion in industry profits.[9] This is happening while the government is still giving billions of dollars in subsidies to the oil and coal companies. Imagine what would happen if the public sector fully and passionately supported the shift to clean, renewable power—and gave those supports to the next generation of power producers. Also, energy conservation measures are readily available. If the United States slashed per capita emissions to current California levels, it would cut its output to 1.7 billion tons below the targets set by the international Kyoto agreement.[10]

Most clean-energy technologies are not mature enough and current clean-energy companies are not large enough to carry the full load right away. That's why the government needs to immediately launch a massive initiative like the Manhattan Project (which invented nuclear technology) or the Apollo Mission (which put a human being on the moon) to solve the riddles of clean energy and perfect these technologies. But we do have some technologies—from solar panels to wind turbines—that are ready to go. To eliminate the dangers of stagflation and avoid climate catastrophe, we should begin deploying these technologies at the pace of a wartime mobilization.

That option sure beats scraping the bottom of the fossil-fuel bucket for another decade or two. It also beats any of the false solutions represented by corn-based ethanol, nuclear power, "clean" coal, or destroying our protected areas in pursuit of the last drop of oil.

FALSE FUELS: CORN, COAL, AND URANIUM

Government-mandated and -subsidized ethanol from corn will go down in history as the "Iraq war" of environmental solutions: ill-considered, costly, and disastrous. In a world full of hungry people, burning food should be criminally punished—not financially subsidized—by the U.S. government. Instead, Washington has granted subsidies and incentives to big agriculture to divert tons of a staple crop into the gas tanks of Americans.[11] As the supermarket and the gas station both fight to get the same ear of corn, the price of that ear goes up and up. And since corn is used in everything—from feeding chickens, pigs, and beef and dairy cows to sweetening sodas—the prices of nearly all food items are also going up. Today food costs are ballooning across the world, causing food riots and pushing vulnerable populations to the brink of starvation and beyond.[12] We should be storing our corn in the stomachs of the world's children, not in the gas tanks of our SUVs. Corn should be food, not fuel.

Nuclear power is also a false solution. Just as there is only so much coal in the ground, there is only so much uranium down there as well. At some point, the laws of supply and demand will catch up to that nonrenewable energy source too.[13] Additionally, mining uranium is messy, destructive, and potentially hazardous for neighboring communities. Highly toxic nuclear waste poses new dangers today, not the least of which is the frightening possibility of fanatics getting their hands on these dangerous materials. And despite clever attempts at repackaging, nuclear power is not even a good "low-carbon" solution for global warming. Constructing a single power plant requires gigantic amounts of concrete—the creation of which spews tons of carbon into the atmosphere. Besides, most proposed plants won't even come online for decades—long past the window for urgently needed reductions.

"Clean coal" is an oxymoron at this point; the technology for it does not even exist. It is a just a great slogan that conceals the awful fact that the mining and burning of coal are two of the dirtiest activities occurring in the United States. "Clean coal" represents a breakthrough in the marketing of coal, but not in the science of burning coal. Proposals for horizontal smokestacks lined with carbon-eating algae do have some appeal,[14] but such facilities exist only in theory at this point. Each power plant would gobble up tons of freshwater and multiple acres of land. And in cooler climates, they would not work at all. We cannot base our entire energy strategy on this idea. Another notion that also should be eliminated as any kind of magic cure-all is the idea that we can pump all of the carbon dioxide from coal-fired power plants into big holes in the ground. It's simply not an option for the vast majority of the world's power plants; few places on Earth have the right geological features underground to even try the experiment. And even then, no container is guaranteed forever.[15]

Besides, even if we can solve the carbon problem for coal, it is still a nonrenewable resource. At some point, coal supplies will drop, prices will rise, and we will be in the same stagflation mess again. Drilling for more oil is no solution either. Prices might dip for a while, but the underlying laws of economics would push them back up. And then what would we do?

Apparently, the fossil-fuel industry's strategy is to convince the American people that we should just burn all the way through the last of our existing oil and coal reserves. Then we can let our freezing, stranded children figure out how to heat their homes and power their vehicles. This is no solution. At best, we will prolong the problem, not solve it. At some point, inevitably, fossil fuels will run out, and we will have to either power down or switch to something else.

Given the climate crisis, we need urgently to begin the transition now. Fossil fuels are a finite resource doing infinite damage. As long as we rely on fossil fuels to power our society, our economy is at risk for stagflation—and our planetary home is at risk too.

Ironically, that's where the good news begins. The generations living today get to retrofit, reboot, and reenergize a nation. We get to rescue and reinvent the U.S. economy. We may as well do it right the first time. We may as well move the society as dramatically as we can in the direction of a fully clean and renewable system. The more aggressive we are, the better off we will be. There is a better future out there.

GREEN-COLLAR JOBS: THE JOBS OF TOMORROW—TODAY!

Some of the barriers to a real breakthrough are not technological, economic, or political. Even under ideal circumstances, rarely discussed practical barriers could prevent us from implementing the economic and energy solutions that are ready at hand.

For example, let's just imagine for a moment that the U.S. president has just signed the best legislation to reverse global warming that anybody ever imagined. Everyone would be clapping and cheering; supporters at the Rose Garden signing ceremony would be weeping with joy. But tomorrow morning, the president of the United States is not going to go out and put up one solar panel. The senators and representatives who passed the law are not going to go out and weatherize one building or manufacture parts for one wind turbine.

So who will do the hard and noble work of actually building the green economy? The answer: millions of ordinary people, many of whom do not have good jobs right now. According to the National Renewable Energy Lab, the major barriers to a more rapid adoption of renewable energy and energy efficiency are not financial, legal, technical, or ideological. One big problem is simply that green employers can't find enough trained, green-collar workers to do all the jobs.[16]

That is good news for people who are being thrown out of work in the present recession. That is good news for people in urban and rural communities who are suffering from chronic lack of work. That is good news for our veterans coming home from Iraq and Afghanistan. That is good news for people returning home from prison, looking for a second chance. And those opportunities for work and wealth creation can be available to all of them—starting right now. Not twenty years from now. Today.

When commentators evoke the "future green economy" or the "green jobs of the future," our minds sometimes start conjuring up images at the far edge of our imaginations. Perhaps we envision a top-secret California laboratory, where strange and mysterious geniuses are designing space-age technologies to save the world. We see cool and beautiful Ph.D.s wearing fancy goggles and green lab coats, turning the dials on strange and wonderful machines. Perhaps someone in the corner is reworking the equations for a new hydrogen fuel cell—or maybe even nuclear fusion. Or maybe we see a courageous space cowboy in orbit, bravely constructing the solar panels that somehow beam down energy to our cities. The possibilities are endless. Someone says "green jobs," and our minds go to Buck Rogers.

Let's be clear: the main piece of technology in the green economy is a caulking gun. Hundreds of thousands of green-collar jobs will be weatherizing and energy-retrofitting every building in the United States. Buildings with leaky windows, ill-fitting doors, poor insulation, outdated furnaces, and old appliances can gobble up 30 percent more energy. That means owners are paying 30 percent more on their heating bills. And it often means that 30 percent more coal-fired carbon is going into the atmosphere. Drafty buildings create broke, chilly people—and an overheated planet.

Another bit of high-tech green technology is the clipboard. That tool is used by energy auditors as they point out energy-saving opportunities to homeowners and renters. This job does not require much training and can be an early entry point into the booming world of energy consultation and efficiency. And one consultation can save an owner hundreds—or even thousands—of dollars annually.

Other green-collar workers can then follow up with other tasks for building owners: wrapping hot-water heaters with blankets, blowing insulation, plugging holes, repairing cracks, hauling out old appliances, installing new boilers or furnaces, replacing old windows with the double-glazed kind. Other pieces of green tech are ladders, wrenches, hammers, tool belts, and nonslip work boots. Those are the "space-age gadgets" used by solar-panel installers every day.[17]

The point is this. When you think about the emerging green economy, don't think of George Jetson with a jet pack. Think of Joe Sixpack with a hard hat and lunch bucket, sleeves rolled up, going off to fix America. Think of Rosie the Riveter, manufacturing parts for hybrid buses or wind turbines. Those images will represent the true face of a green-collar America.

If we are going to beat global warming, we are going to have to weatherize millions of buildings, install millions of solar panels, manufacture millions of wind-turbine parts, plant and care for millions of trees, build millions of plug-in hybrid vehicles, and construct thousands of solar farms, wind farms, and wave farms. That will require thousands of contracts and millions of jobs—producing billions of dollars of economic stimulus.

And don't think of green-collar workers as laboring only in the energy sector. Though the need for a clean-energy revolution will be the main driver in revamping the economy, we will also need well-trained, well-paid workers in a range of green industries: materials reuse and recycling, water management, local and organic food production, mass transportation, and more.

We will have to completely overhaul not just the economy, but the way we think about the economy. The foundations for most of our economic models, accounting tools, and business practices have their roots in the eighteenth and nineteenth centuries. At that time, there was an awful lot of nature—and relatively few people. Today there are an awful lot of people, but shockingly little nature left. Most Western economic models assume cheap and abundant energy forever. They assume cheap and abundant everything forever—such that we can throw millions of tons of materials into landfills and in-cinerators every year, year after year, and never run out of anything important.

However, the price signals alerting us to this folly are starting to kick in. Today we live on what author and *New York Times* columnist Tom Friedman calls a hot, flat, and crowded planet.[18] That means there are appreciable and obvious limits to resources that the textbooks once described as inexhaustible. Over the course of this century—out of choice, necessity, or both—we will rework our economic and business models to reflect that reality.

In the meantime, opportunities abound to make things better for everyone. It is not as if the present economy has been so perfect that everyone should cling to it—fingers and toes—for fear of any changes. The U.S. economy and society have been malfunctioning for some time. Green-collar jobs could help us conserve resources, create new sources of energy, and give the nation the power to grow the economy again. What's more, we have the chance to build this new energy economy in ways that reflect our deepest values of inclusion, diversity, and equal opportunity for everyone.

The key to this is setting high standards and expectations for what a green-collar job even is. That starts by baking high quality and good values into the very definition of a green-collar job. My definition of a green-collar job is this: *a family-supporting, career-track job that directly contributes to preserving or enhancing environmental quality.* Like traditional blue-collar jobs, green-collar jobs range from low-skill, entry-level positions to high-skill, higher-paid jobs and include opportunities for advancement in both skills and wages. Think of them as the 2.0 version of old-fashioned blue-collar jobs, upgraded to respect the Earth and meet the environmental challenges of today.

Green-collar jobs can and should be good jobs. Like blue-collar jobs, green-collar jobs can pay family wages and provide opportunities for advancement along a career track of increasing skills and pay. We should never consider a job that does something for the planet and little to nothing for the people or the economy as fitting the definition of a green-collar job. A worthwhile, viable, and sustainable green economy cannot be built with solar sweatshops.

Here's more good news. Most green-collar jobs are middle-skill jobs. That means they require more education than a high-school diploma, but less than a four-year degree. So these jobs are well within reach for lower-skilled and low-income workers, as long as they have access to effective training programs and appropriate supports. We must ensure that all green-collar-job strategies provide opportunities for low-income people to take the first step on a pathway to economic self-sufficiency and prosperity.

The green economy demands workers with new skill sets. Some green-collar jobs—say, renewable-energy technicians—are brand-new. But others are actually existing job categories that are being transformed as industries transition to a clean-energy economy, for example, computer control operators who can cut steel for wind towers as well as for submarines or mechanics who can fix an electric as well as an internal combustion engine. We can identify the specific skills the green economy demands. Then we can invest in creating new training programs and retooling existing training programs to meet the demand.

Even better news. Much of the work we have to do to green our economy involves transforming the places we live and work in and changing the way we get around. These jobs are difficult or impossible to outsource. For instance, you can't pick up a house, send it to China to have solar panels installed, and have it shipped back. In addition, one major group of manufacturing jobs—a sector that has been extensively outsourced—is producing component parts for wind towers and turbines. Because of their size and related high transportation costs, they are most cost-effectively produced as near as possible to wind-farm sites. Cities and communities should begin thinking now about ways their green strategies can also create local jobs.

Both urban America and rural America have been negatively impacted over the past decades by a failure to invest in their growth. Green-collar jobs provide an opportunity to reclaim these areas for the benefit of local residents. From new transit spending and energy audits in inner cities to windmills and biomass operations in our nation's heartland, green jobs mean a reinvestment in the communities hardest hit in recent decades.

The "green" in "green-collar" is about preserving and enhancing environmental quality—literally saving the Earth. Green-collar jobs are in the growing industries that are helping us kick the oil habit, curb greenhouse-gas emissions, eliminate toxins, and protect natural systems. Today, green-collar workers are installing solar panels, retrofitting buildings to make them more efficient, refining waste oil into biodiesel, erecting wind farms, repairing hybrid cars, building green rooftops, planting trees, constructing transit lines, and so much more.

The green economy should not be just about reclaiming thrown-away stuff. It should be about reclaiming thrown-away communities. It should not just be about recycling materials to give things a second life. We should also be gathering up people and giving them a second chance. Formerly incarcerated people deserve a second shot at life—and all obstacles to their being able to find that second chance in the green sector should be removed. Also, our urban youth deserve the opportunity to be a part of something promising. Across this nation, let's honor the cry of youth in Oakland, California, for "green jobs, not jails."

In other words, we should use the transition to a better energy strategy as an opportunity to create a better economy and a better country all around. In fact, we should see this whole process as a "break-up" situation. When you break up with your lover, it is tough at first. But the next weekend you start going to the gym, you quit smoking, you buy some new clothes. You can use the energy unleashed by one big change to positively transform your life for the better. Well, we in America are about to break up with oil. Why not break up with poverty and discrimination too?

If we decide to do that, we can do something extraordinary. We can connect the people who most need work to the work that most needs to be done—we can fight pollution and poverty at the same time.

We have the chance now to create new markets, new technology, new industries, and a new workforce. Let's do it right—with good wages, equal opportunity, and pathways to success for those whom the pollution-based economy left behind.

We had the opportunity to do all of this decades ago, but we blew it. In the 1970s energy crisis, U.S. president Jimmy Carter and California governor Jerry Brown gave the go-ahead to very aggressive programs in alternative energy.[19] Practically every major breakthrough we associate with clean energy was created or dramatically improved during that short window of time—from solar photovoltaic to wind turbines.[20] For a while, we even had solar panels on the White House. But President Ronald Reagan took them down. Soon gas prices dropped, and we relapsed completely back into our oil addiction. Other countries—namely, Germany and Japan—took our technological breakthroughs and turned them into major economic boons for themselves. Meanwhile, we have fallen behind on the clean-energy revolution that we initiated. We started down the right road, but we turned back. It hurts to think how much stronger our economy and our planet would be today, had we continued full bore in the direction we were headed thirty years ago.

Now it is time to head back down the right road again. If the federal government shifts its policy to fully back the green economy, the private sector can create millions more jobs in new clean and green industries. A smart climate bill—which mandates reductions in greenhouse gases and ends our oil dependence—would not wreck the economy, but save it. It would eventually break the back of stagflation. It would be an economic stimulus package on steroids.

But no single group can win that monumental victory in Washington by itself. Certainly, affluent, mostly white environmental lobbying groups cannot win a comprehensive victory on their own. No top-down agenda, dictated to society by a perceived eco-elite, will win acceptance in a country as proud and diverse as ours. Any kind of elitist approach will fuel resentment and generate a "backlash alliance" between polluters and poor people. For the sake of the planet, the effort to green the economy must be owned by a much bigger and broader coalition. Most environmental groups know this, but adjusting to true partnership with very different people may not be easy for some of them.

There is reason to worry. Here we must face honestly some hard and painful truths about a sad history of racial blind spots and class exclusion in some parts of the mainstream environmental movement. I raise the point not to pick at old scabs, but to help avoid new injuries. The truth is that even the greatest ecological victories— from the Wilderness Act protecting the land to the establishment of the Superfund to clean up hazardous waste sites—could have been better engineered to involve, include, and help more people. As we build this new green wave, the new environmentalists will need to work in partnership with people of all classes and colors—not just because it is the right thing to do, but because it is also the best way to ensure that we are doing things right.

And yet Velcro needs two sides to stick. Working-class people, people of color, religious leaders and groups, and other nontraditional constituencies also need to step up to the plate. The time has come for us to stop letting a small number of groups and leaders carry the load on environment and energy policy. Rather than complaining about the way they do it, all of us must step forward, take responsibility, and say in our own voices: "This is our Earth too. We are going to be a part of saving it."

If we do that—and if a critical mass of people from all races, faiths, genders, and classes were to see the mutual benefit of cocreating the future economy—a visible and dramatic transformation of the entire economy is entirely possible. And I am hopeful that a new kind of movement will emerge from our joint efforts to save the planet and the people.

Let us all say together: "We want to build a green economy strong enough to lift people out of poverty. We want to create green pathways out of poverty and into great careers for America's children. We want this 'green wave' to lift all boats. This country can save the polar bears and poor kids too."

Let us say: "In the wake of Katrina, we reject the idea of 'free market' evacuation plans. Families should not be left behind to drown because they lack a functioning car or a credit card. Katrina's survivors still need our help. And we need a plan to rescue everybody next time. In an age of floods, we reject the ideology that says we must let our neighbors 'sink or swim.'"

Let us say: "We want to ensure that those communities that were locked out of the last century's pollution-based economy will be locked into the new clean and green economy. We know that we don't have any throwaway species or resources, and we know that we don't have any throwaway children or neighborhoods either. All of creation is precious and sacred. And we are all in this together."

Those words would open the door to a cross-race and cross-class partnership that would change America and the world. The idea of a new "social-uplift environmentalism" could serve as the cornerstone for an unprecedented "Green Growth Alliance." Imagine a coalition that unites the best labor and business leaders, social justice activists, environmentalists, intellectuals, students, and more—all sharing the burdens and benefits, risks and rewards, of the journey to a green-collar economy. The power of that combination would rival the last century's most powerful alliances: the New Deal and New Right coalitions.

Imagine a Green New Deal—with a pivotal role for green entrepreneurs, a strategic and limited role for government, and an honored place for labor and social activists. Such a force would change the direction of our society. It would put the government on the side of the problem solvers in the U.S. economy, not the problem makers—and bring us all together.

The road to unity and ultimate victory will be hard, with many pitfalls. But to understand why such a journey is necessary and vital, we first must understand the true dimensions of the peril we are facing.

ONE

The Dual Crisis

For forty-eight hours, Larry and Lorrie waited for the "imminent" arrival of the buses, spending the last twelve hours standing outside, sharing the limited water, food, and clothes they had with others. Among them were sick people, elders, and newborn babies. The buses never came. Larry later learned that the minute the buses arrived at the city limits, they were commandeered by the military.

Walgreen's remained locked. The dairy display case was clearly visible through the widows. After forty-eight hours without electricity, the milk, yogurt, and cheeses were beginning to spoil in the ninety-degree heat. Without utilities, the owners and managers had locked up the food, water, disposable diapers, and prescriptions and fled the city. Outside, residents and tourists grew increasingly thirsty and hungry. The cops could have broken one small window and distributed the nuts, fruit juices, and bottled water in an organized manner. Instead, they spent hours playing cat and mouse, temporarily chasing away the looters.

Repeatedly, Larry and Lorrie were told that resources, assistance, buses, and the National Guard were pouring into the city. But no one had seen them. What they did see—or heard tell of—were electricians who improvised long extension cords stretching over blocks in order to free cars stuck on rooftop parking lots. Nurses who took over for mechanical ventilators and spent hours manually forcing air into the lungs of unconscious patients to keep them alive. Refinery workers who broke into boatyards, "stealing" boats to rescue people stranded on roofs. And other workers who had lost their homes, but stayed and provided the only assistance available.

By day four, sanitation was dangerously abysmal. Finally Larry and Lorrie encountered the National Guard. Guard personnel said that the city's primary shelter, the Superdome, had become a hell-hole. They also said that the city's only other shelter, the Convention Center, was also descending into chaos and squalor and that the police were not allowing anyone else in. They could offer no alternatives and said, no, they did not have extra water to share.

When Larry and Lorrie reached it, the police command center told them the same thing. Without any other options, they and their growing group of several hundred displaced people decided to stay at the police command post. They began to set up camp outside. In short order, the police commander appeared to address the group. He told the group to walk to the expressway and cross the bridge, where the police had buses lined up to take people out of the city. When Larry pressed the commander to make certain this wasn't further misinformation, the commander turned to the crowd and stated emphatically, "I swear to you that the buses are there."

The group set off for the bridge with great hope and were joined along the way by families with babies in strollers, people using crutches, elderly clasping walkers, and others in wheelchairs. It began to pour down rain, but the group marched on.

As they approached the promised location, they saw armed sheriffs forming a line across the foot of the bridge. Before Larry and

Lorrie were even close enough to address them, the sheriffs began firing their weapons over people's heads. The crowd scattered and fled, but Larry managed to engage some of the sheriffs in conversation. When told about the promises of the police commander, the sheriffs said there were no buses waiting.

Larry and Lorrie asked why they couldn't cross the bridge anyway. There was little traffic on the six-lane highway. The sheriffs refused.

Heartbroken and desperate, the group retreated back down the highway and took shelter from the rain under an overpass. After some debate, they decided to build an encampment on the center divide of the expressway, reasoning that it would be visible to rescuers and the elevated freeway would provide some security. From this vantage point they watched as others attempted to cross the bridge, only to be turned away. Some were chased away with gunfire, others verbally berated and humiliated. Thousands were prevented from evacuating the city on foot.

From a woman with a battery-powered radio they learned that the media were talking about the encampment. Officials were being asked what they were going to do about all those families living up on the freeway. The officials responded that they were going to take care of it. "Taking care of it" had an ominous ring to it.

Sure enough, at dusk a sheriff rolled up in his patrol vehicle, drew his gun, and started screaming, "Get off the fucking freeway!" A helicopter arrived and used the wind from its blades to blow away the flimsy shelters. As Larry and Lorrie's group retreated, the sheriff loaded up his truck with the camp's small amount of food and water.

Forced off the freeway at gunpoint, they sought refuge in an abandoned school bus under the freeway, more terrified of the police and sheriffs with their martial law and shoot-to-kill policies than of the criminals who supposedly were roaming the streets.

Finally a search-and-rescue team transported Larry and Lorrie to the airport, where their remaining rations, which set off the metal detectors, were confiscated. There they waited again, alongside thousands of others, as a massive airlift gradually thinned the crowds and delivered them to other cities across the region.

After they disembarked from the airlift, the humiliation and dehumanization continued. The refugees were packed into buses, driven to a field, and forced to wait for hours to be medically screened to make sure no one was carrying communicable diseases. In the dark, hundreds of people were forced to share two filthy, overflowing porta-potties. Those who had managed to make it out with any possessions were subjected to dog-sniffing searches. No food was provided to the hungry, disoriented, and demoralized survivors.[1]

AMONG THOSE LEFT behind after Katrina, they were the lucky ones. Larry and Lorrie are a Caucasian couple who had some resources available to them. The whole world knows what happened to the poor, black residents of New Orleans who had none.

I believe stories like this deserve retelling, revisiting, remembering. Stories from Katrina's aftermath demonstrate that the issues of poverty, climate destabilization, petrochemical poisons, and the vulnerabilities of an oil-based economy are not just petty obsessions of the politically correct crowd. They are life-and-death issues for real people.

To be clear, it wasn't Hurricane Katrina that wrought that catastrophe. It was a "perfect storm" of a different kind: neglect of our national infrastructure combined with runaway global warming and blatant disregard for the poor.

The flooding was not a result of heavy rains. It was a result of a weak levee—one that was in mid-repair when the storm hit. And that levee collapsed for one simple reason: fixing it was not a priority for our country's administration. Instead, funds that should have gone for our infrastructure and to the repair of the levee were allocated to the war effort.

The dollars that could have saved New Orleans were used to wage war in Iraq instead, a war undeniably linked to our dependency on that region's oil. Additional funds that might have spared the poor in New Orleans and the Gulf region (had the dollars been properly invested in levees and modern pumping stations) were instead passed out to the rich as tax breaks. The Katrina disaster and what followed clearly point to the fact that overfunding the military and cutting services make us *less*, not more, safe.

Yet that is only one of the lessons. As Pulitzer Prize–winning journalist Ross Gelbspan said: "Katrina began as a relatively small hurricane that glanced off south Florida, [but] it was supercharged with extraordinary intensity by the relatively blistering sea surface temperatures in the Gulf of Mexico."[2] In other words, global warming supercharged the hurricane. Yet American energy policies continue to add even more carbon to the atmosphere, further destabilizing the climate.

The human suffering in the floodwaters was not—and continues not to be—equally distributed. Poor people and black people didn't "choose to stay behind." They were left behind. The evacuation plans required the city's residents to have working private cars—plus gas money and nearby relatives or funds for a hotel stay. If people didn't have those things, tough luck.

Had the responsible agencies valued the lives of the poor, they would have helped the destitute flee in the face of the hurricane—even those who couldn't afford a car or a motel room. But when the face of suffering is mostly black, somehow our high standards for effective action and compassion begin to sag.

The story of Katrina is a rare political circumstance, a genuine teaching moment. We owe it to the dead not to waste it. We cannot allow a messy stew of shame, pain, and racial disdain to prevent us from looking deeply into the heart of this disaster. If we can look into it and not turn away, we learn something invaluable: that we are all living in a floodplain today.

For some of us this is literally the case, like New Orleans's Ninth Ward residents. Poverty has forced many people into homes in neighborhoods that are vulnerable to everything from flooding to mudslides to toxic air—as if it isn't destabilizing enough to have to worry about safety on war-torn streets, get an education in schools with no resources, or hunt for scarce jobs.

Meanwhile, the stability of the relatively affluent is also under threat. The average American family has spent itself out onto a perilous perch. Credit-card debt outstrips savings plans. A sharp economic downturn or the collapse of the U.S. dollar could toss millions overboard into financial crisis.

And of course we are also on the verge of environmental bankruptcy. That big greenhouse-gas bill is fast coming due—in the form of extreme weather events that could overtake more than just the Gulf Coast. Some say it could be Manhattan, and most of our cities are no more ready than New Orleans was. Our levees, dams, schools, and hospitals are crumbling or in poor repair.

On a larger scale, Katrina also shows the flaws of the individualist "sink or swim" philosophy that dominates both major political parties. That political-economic worldview informed New Orleans's free-market evacuation plan, which ensured that only those with private cars and money could get out.

The Katrina story illustrates clearly the two crises we face in the United States: radical socioeconomic inequality and rampant environmental destruction.

CRISIS #1:
RADICAL SOCIOECONOMIC INEQUALITY

Given the skyrocketing energy prices and the specter of stagflation, it will be hard to revive the sputtering U.S. economy. But even before the present energy crisis and economic downturn, the U.S. economy and society were in deep trouble. We will examine in some detail the symptoms of a grave malady.

The country has long been deep in the throes of a socioeconomic crisis, one characterized by contracting economic opportunity for working people, growing disparities between the races, and the hording of immense wealth and privileges at the very top of our society. These features are getting worse, not better.

In fact, the United States is experiencing the greatest economic inequality between its wealthiest and poorest citizens since the Great Depression of the 1930s.[3] While a tiny number of people at the top amass wealth, very little is left for everyone else to get by on; today more than 34 percent of the country's private wealth is held by just the richest 1 percent of people. Their wealth equals more than the combined wealth of the bottom 90 percent of people in this country.[4]

In comparison to their employees, chief executive officers of major corporations are earning ungodly sums of money. Their salaries are four hundred times higher than that of the average worker. That outrageous disparity has been growing exponentially: in 1990 executive pay was (just?!) about a hundred times that of the average worker.[5] Meanwhile Americans are working longer, if not harder, than ever: we have added eighty hours to our work year over the last twenty-five years.[6]

Meanwhile, 15.6 million American households live in extreme poverty (their incomes are below half the amount considered the poverty line, which was $20,650 for a family of four in 2007), the highest rate recorded since researchers started tracking those numbers in 1975.[7] Since 2000, the country has lost more than 3 million manufacturing jobs.[8] More than 44 million of us live without health insurance—a number that continues to grow.[9] In too many homes, the family health plan is short and simple: "Nobody get sick!" At the same time, the great majority of Americans are less and less able to get out of debt or to save money for the rainy days that are coming.

So many folks have needed to declare personal bankruptcy that the government finally decided to step in.[10] Unfortunately, the government did not move to bail people out (as the Fed did for Bear Stearns investment bank when it got in trouble in 2008).[11] To the contrary, Congress saw the tidal wave of bankruptcy petitions as a sign of threat—not to families, but to the profits of the credit-card industry. So it changed the laws to make it even *harder* for card holders to declare bankruptcy and free themselves from usurious interest rates and outrageous fees.

As painful as it is to acknowledge, factors of race and gender exacerbate the inequities. Even today, female workers earn 77 cents for each dollar earned by their male counterparts.[12] People of color own a mere 18 cents for every dollar of white wealth.[13] Median income levels are lowest among females of all races, and significantly lower for black, Latino, or Hispanic women.[14] And income levels are related to educational attainment—whether someone graduates from high school or attends college: compared to the 32 percent of whites who hold a bachelor's degree, only 19 percent of blacks and 13 percent of Hispanics hold one.[15] About 33 percent of African American children, 29 percent of Native American children, and 28 percent of Latino children live below the poverty line, compared to 9.5 percent of white children.[16] In our school system, students of color often are more likely to be taught by an underqualified teacher; they are nearly twice as likely as white students to attend overcrowded schools.[17]

New immigrants and undocumented workers suffer unequal and unfair treatment. Our society profits from the labor of 11 million people[18]—many of whom pick our food, nurse our children, clean up after us—without embracing them fully, without honoring their work, and without extending to them the same rights and respect we would demand for ourselves.

In the area of health and wellness, we see race-correlated differences as well. Infant mortality among black Americans is more than twice the rate for whites. Rates of cancer are 25 percent higher for blacks than for whites. Asthma death rates are more than twice as high for Latinos than for whites. Deaths from diabetes are more than twice as high for blacks than for whites. Tuberculosis and hepatitis B also show up more in communities of color.[19]

Disparities are concentrated in some places. Over the past decades, manufacturing jobs in cities have disappeared. As a result, employment opportunities for young, unskilled men living in inner cities eroded significantly through the 1960s and 1970s. At the same time, anxious whites and affluent climbers fled to the suburbs. In some inner-city neighborhoods, basic services deteriorated—including health care, stores, schools, garbage collection, police and fire protection, and employment options. Poverty, social chaos, and violence ensued. Today the life expectancy for African American men living in places like Harlem and Washington, D.C., (57.9 years) is lower than for men in Bangladesh (58.1) and Ghana (58.3).[20]

In the midst of all that suffering, the incarceration industry saw and seized a huge growth opportunity. In today's criminal justice system, the racial disparity is astounding. The Drug Policy Alliance reports:

Although African Americans comprise only 12.2 percent of the population and 13 percent of drug users, they make up 38 percent of those arrested for drug offenses and 59 percent of those convicted of drug offenses, causing critics to call the war on drugs the "New Jim Crow." The higher arrest rates for African Americans and Latinos do not reflect a higher abuse rate in these communities but rather a law enforcement emphasis on inner-city areas where drug use and sales are more likely to take place in open-air drug markets where treatment resources are scarce.[21]

As a result of these and other disparities, African Americans are seven times more likely to go to jail than whites. Although youth of color represent one-third of the adolescent population in this country, they represent two-thirds of our country's juvenile inmates.[22]

Just to be clear, people of color and the urban poor are not the only ones hurting. In many ways, those suffering in rural and small-town America are even worse off, because their plight doesn't even make the sensationalized evening news. Midsize family farmers are being squeezed—they are struggling to get fair prices for their products and compete on the open market as farm consolidation pushes more of them out of business. High fuel prices are hitting them especially hard. Meanwhile, they have less access to quality doctors; rural schools have a tough time recruiting teachers.

Those small towns that are able to attract businesses often find themselves being converted within a few years into an indistinguishable, cookie-cutter version of every other town in the country. Big-name chains, strip malls, and big-box giants are replacing the mom-and-pop stores that once gave such life and distinctive character to America's hometowns.

Furthermore, the U.S. military draws a disproportionate share of its recruits from low-income families in economically distressed parts of the country.[23] Soon, thousands of young veterans will be coming home from Iraq and Afghanistan. They will have injuries, both visible and invisible, but most will have no visible job prospects. If there is going to be a functioning society into which they can reintegrate, we must work to create a national economy that works better for everyone.

That means we have our work cut out for us. Wages, wealth, health, homes, schooling, fairness in the courts, youth opportunity: the low performance and disparities in these categories are both shocking and shameful. This is a nation that has committed itself to equal protection, equal opportunity, and "liberty and justice for all." We are earning a failing grade on those core values. The totality of these negative outcomes cannot be dismissed as just a blip on the screen or passing problem. They represent a serious system failure— one that we must muster the will and the courage to correct.

But we must do so without worsening the other major crisis we face: the ecological crisis, epitomized by the climate catastrophe that our present economy is courting.

CRISIS #2:
RAMPANT ENVIRONMENTAL DESTRUCTION

I won't take up a lot of space and time attempting to convince you of the reality of climate change and global warming or of human responsibility for these events. There are many excellent books and Web sites that describe the consensus of the most significant and well-respected scientific institutions, including the Intergovernmental Panel on Climate Change (IPCC), the American Meteorological Society, the American Geophysical Union, and the American Association for the Advancement of Science.

Their predictions grow increasingly clear—and frightening—as they compare vast amounts of climate and natural resource–related data and compare them to historical records of temperatures and conditions on Earth. An overwhelming number of experts also agree that human consumption of natural resources and our treatment of the

Earth are largely responsible.[24] It is no longer a question of whether or not climate change is happening; it is a matter of how soon—and how hard—it will hit. When it comes to the looming climate crisis, everything is a question of degrees.

By burning fossil fuels to meet our ravenous hunger for power in our homes, factories, and means of transportation, humanity adds about seven billion tons of carbon (twenty-six billion tons of carbon dioxide) to the atmosphere every year.[25] Meanwhile we keep chopping down trees; those trees are the lungs of the planet, pulling carbon out of the air and breathing out oxygen. Therefore, clear-cutting whole continents undermines the Earth's overall ability to soak up the carbon dioxide. In effect, we are running the carbon faucets at full blast, while we plug up the carbon sinks. As a result, our carbon cup runneth over.

That is why temperatures are rising at unprecedented rates. Over the course of the twentieth century, mean temperatures rose by 1.4 degrees Fahrenheit (.8 degree centigrade). But global temperatures have been quickly accelerating since the 1980s, with the top ten warmest years on record occurring since 1990. The IPCC employs approximately two thousand scientists working on climate models and predictions; their frightening projections plot a temperature increase of 10.6 degrees F (5.9 C) by the end of the century if greenhouse-gas emissions continue unabated.[26]

The temperature shifts have already set in motion many mostly destructive weather changes. Depending on the region, warming causes drought, floods, blizzards, cyclones, and other extreme weather. Deserts are expanding inexorably. Fully one-third of the Earth's land, or seven times today's percentage, according to the British Meteorological Office, will be subject to extreme drought by the end of the century.[27]

The oceans are already rising, because warmer water expands. But if the Greenland ice sheet melts, then untold numbers of coastal settlements, low-lying islands, and croplands will be submerged. All totaled, climate-driven water disasters—storms, floods, and droughts—claimed more than half a million lives between 1991 and 2000, costing the global economy the equivalent of fifty dollars for every human on Earth.[28]

As precious frozen supplies of freshwater melt into the salty seas and rivers dry up in the heat, the demand for water will further outstrip supply. The World Bank predicts that two-thirds of the global population will suffer from lack of access to freshwater by 2025.

Absorption of carbon dioxide has altered the pH balance of the oceans, resulting in the possible extinction of many shell-forming organisms like corals, which in turn impacts the aquatic species that depend upon them. And those are far from the only species we have to be concerned about losing. The most sensitive species—those with narrow temperature tolerances limited to extreme cold (like our friends the polar bears) or tropical climates—are going extinct as their habitats become inhospitable. The IPCC estimates that 30–40 percent of all species are at risk of extinction with projected levels of warming.[29]

Yet warmer temperatures are more hospitable to some. Insects that bring diseases such as malaria, dengue fever, and possibly new tropical viruses thrive. So do insect pests like locusts and corn borers, which can spread to higher altitudes and latitudes, decimating more crops in longer active periods as the winters get milder. Fungal and bacterial plant diseases flourish too. That's life on a superheated planet: if the floodwaters don't get you, the pestilence and plagues will.

Scientists estimate that ecosystems can adapt to a temperature change of only 1.8 degrees F (1 C) over a century. More than that, and the changes in habitat, mass extinctions of species, and disturbances to the predator/prey balance throw off our entire food chain.[30] Some of our key cereal crops fail with increased carbon dioxide levels. But aggressive weed varieties love the stuff. And that's not the worst news for farmers or people who like to eat. Water shortages will make growing food even harder. Topsoil—already badly eroded—loses nutrients even faster at higher temperatures and is more prone to being washed away as it dries out and becomes compacted.

The British Meteorological Office predicts that fifty million additional people will be starving or facing severe food shortages by 2050. Others estimate that, on a hotter planet, *hundreds* of millions will starve.[31] Particularly in Africa, which is considered one of the places most vulnerable to climate change, millions of people will face food and water shortages as early as 2020.[32]

Catastrophic events disproportionately impact the poor. Certainly weather instability and other environmental changes like rising ocean levels can hit any part of the country or the world, but the poor usually have fewer resources to protect themselves. And it is especially ironic—and horribly unfair—that the people who are least blameworthy are likely to suffer the most. Those of us in industrialized countries generate sixty times more carbon dioxide pollution per person than people in the least industrialized countries.[33] So it is *our* lifestyle that is most to blame for the coming troubles.

But the guiltless will bear the brunt and suffer the wrath of an enraged Mother Nature. People in wealthy countries can cushion the blows. That's why, as the 1999 edition of the World Disasters Report concludes, about 96 percent of all deaths from natural disasters happen in developing countries.[34] Our actions—and refusals to act—in the wealthier nations are funneling more disasters and death toward the poorest people on Earth.

Some reputable experts have already given up hope. They say it is already too late and that too much damage has already been done to the climate to avert catastrophe. But there is still widely held scientific opinion that if we radically cut our greenhouse-gas emissions, we can keep warming below 3.6 degrees F (2 C). And at that level, we can avert or manage the worst of the damage to our ecosystem and our lives. In large part, cutting greenhouse-gas emissions means developing clean energy and increasing efficiency.

To achieve the needed reductions, we will need both political and economic transformation—immediately. The necessary solution is to establish the kind of politics and policies that could win over a critical mass of U.S. citizens and inspire them to launch a crash program in conservation and renewable energy—so that we can save our ability to survive on the only planetary home we have ever known.

No matter what we do, however, we can be sure that the economy and the environment will both get worse before they get better. That is why this chapter began with the story of the time that a storm came and this nation left its most vulnerable people—its poor, black, elderly, and disabled people—behind to die. We must sear the moral of that story into the memory of this nation. This catastrophe—and its lessons—must become part of our national legend. Only then can we be assured that the mind-set that permitted it will never again be allowed to lead this country.

Even today, New Orleans still lies demolished. The survivors are scattered; some of them are still living in the toxic trailers provided by the Federal Emergency Management Agency (FEMA). Many sections of the impoverished Ninth Ward still look exactly as they did after the waters receded. The devastation serves as a reminder of the magnitude of the problems we face—both social and ecological.

Is there a way to address both crises simultaneously? Can we help the people without harming the planet? Can we protect the planet, without dooming more people to material poverty?

I believe the answer is yes. And if so, the key to a dual victory is to be found in the heart of the one sector of the U.S. economy that is still thriving and growing: the green part.

The Fourth Quadrant

IN THE LAST chapter, we examined a lot of discouraging data. However, as the picture becomes fuller and clearer, we will see how we can restore confidence and meet the many challenges before us.

One powerful source of hope is the explosion of interest in this country in all things "green." Now, just as the ecological crisis nears the boiling point, our society is entering the third wave of environmentalism. And just in time. The key to solving both the economic and the ecological problems described in the last chapter can be found in the emerging green wave.

Environmentalism's first wave was the "conservation" movement of the early 1900s, which worked to preserve and conserve the best of the natural past. The second wave was the "regulation" wave of the late 1960s and 1970s, which sought to manage the problems of the industrial present. Both of these waves continue to this day; in fact, their work is more necessary than ever.

Now a new wave is emerging alongside them, and it is a fundamentally different phenomenon. It is not focused on saving the beauty and bounty of the past, as important as that work is. Nor is it focused on regulating the problems of the present, as vital as that work is. The new stage is focused, instead, on inventing solutions for the future. The new green wave is an "investment" wave, and it has the power to change the world.

We will trace the origins and dynamics of all three phases. Then we will explore the profound importance of the recent explosion in ecological invention, entrepreneurship, and investment. And I make the case that the entire future of U.S. politics and economics will be bound up with the direction and final fate of the new green wave.

Before we look too far ahead, it is wise to ground ourselves by looking back.

ENVIRONMENTALISM'S FIRST WAVE: CONSERVATION

There was a time when America didn't need an environmental movement. The original human inhabitants of this land, the Native Americans, were geniuses at living in harmonic balance with their sister and brother species. Before Europeans arrived, the entire continent was effectively a gigantic nature preserve. Squirrels could climb a tree at the Atlantic Ocean and move branch to branch to branch until they reached the Mississippi River. So many birds flew south for the winter that their beating wings sounded like thunder; their numbers blotted out the sun. And the Native Americans achieved this feat of land management and sustainability over thousands of years, on a continent that was fully populated by humans.

How did they manage it? The worldviews of indigenous peoples connected them inextricably to the cosmos, the climate, the land—and its plant and animal inhabitants. It still does. Indian's economic and social structures were and are based on their knowledge and understanding of natural processes and cycles like the tides, the phases of the moon, the movements of stars, the seasons, and the reproductive cycles and migrations of animals. In this worldview, nothing is linear; everything is cyclical.

As a result, Native peoples practice reciprocity. They honor the gifts of the Earth and its creatures and give back in return for them. As a basket weaver from the California Pomo tribe instructed, "When you come to dig these basket roots, you don't rush in there and run all over. You don't do that. My mother always approached this grass very slowly. She'd come and stand and say a prayer. . . . She always asked the Spirit to give her plenty of roots. Then she'd say 'Thank you, Father,' before she dug. And after she'd finished and had got what she wanted, she said a prayer, which is like saying 'That's good, you gave me enough. Amen, Father.'"[1] Another tradition might call it "mindfulness."

Perhaps most important, Native Americans traditionally believe in accountability. The Great Law of the Iroquois Confederacy states: "In our every deliberation, we must consider the impact of our decisions on the next seven generations." We humans, being just one component of an interconnected system that also extends forwards and backwards in the continuum of time, inherit a world shaped by the actions of our forbears. And we are responsible for holding that world in trust for all the generations to come.

Environmental stewardship wasn't the only area in which early Native Americans excelled. In contrast to a prevailing image of indigenous peoples as primitive or less evolved, paleoanthropologists have found that tribal peoples in the past had stronger bones, lower rates of infant mortality, and fewer dental cavities and signs of degenerative illnesses than today's "civilized" and immunized people.[2] The leading American indigenous civilizations achieved world-historic heights of political statesmanship; an example is the founding of the Iroquois Federation, a model for the framers of the U.S. Constitution. It is a tragedy of the highest order that

Native legacy and wisdom were almost entirely decimated—along with the environment—when their lands came under new management in 1492.

The European colonizers "discovered" what was to them a "new world," a land of unimaginable beauty and splendor. But they exploited it relentlessly, razing acres of ancient and majestic trees, wiping out whole populations of gullible seals and buffalo, and killing entire societies of indigenous people. In the little more than a hundred years between 1769 and 1890, for example, the population of California Indians dropped from 310,000 (conservatively estimated) to 17,000.[3] The Europeans grew nonnative plants; they hunted animals not to subsist, but to sell the meat or the furs. Their enterprises wreaked havoc on the native ecological systems. The European conquest drove countless species into extinction.[4]

These are hard and ugly facts, but they are a part of the nation's legacy. The sad truth is that the world's most powerful democracy was founded on land stolen from Native Americans. And it was built, in surprisingly large measure, by labor stolen from enslaved Africans. These are two of the birth defects of the republic, and we forget them at our own peril. They set a low mark from which every generation of Americans must actively strive to further distance the country—through repairing the Earth, opposing racism, respecting treaty obligations, and uplifting the disadvantaged of all colors. It is the patriotic duty of all who enjoy this society's bountiful fruits to also face up to its painful roots—and to take intelligent steps, in each generation, to find ways to make things better.

In accepting this responsibility, one places oneself in good company. After all, a movement of white abolitionists eventually rose up to oppose slavery. And the children of European settlers eventually did rise up to defend—if not the red inhabitants of the land—then at least the land itself. Their noblest efforts deserve our eternal respect. Historians call those defenders of the land "conservationists."

They were far from perfect people, but our quest for hope begins, in many ways, with their struggles to preserve and conserve the bounty and richness of these lands.

During the late 1800s and early 1900s, American writers like Ralph Waldo Emerson, Henry David Thoreau, George Perkins Marsh, and John Muir were extolling the sacred qualities and beginning to note the fragility of the natural world. Their words stirred concern in middle-class Americans, influencing public opinion and, ultimately, government policy. Theodore Roosevelt, the twenty-sixth president of the United States, is generally acknowledged as the first president to have prioritized the conservation of natural resources. Rightly so. During his tenure, he spearheaded and accomplished— often by executive fiat—a great number of landmark environmental conservation victories.[5]

What motivated his actions? As it turns out, Roosevelt was an unhealthy youngster. He wrote of his childhood in his *Autobiography:* "I was a sickly, delicate boy, suffered much from asthma, and frequently had to be taken away on trips to find a place where I could breathe."[6] The places he breathed freely were unpaved and green, and it seems likely that he never entirely forgot the debt he owed nature for rescuing his lungs.

Roosevelt said, "The relation of the conservation of natural resources to the problems of National welfare and National efficiency had not yet dawned" before he took office, and no comprehensive data on the scope or condition of the nation's resources existed.[7] In response, Roosevelt established the National Conservation Commission, which completed the first inventory of national resources.[8]

In the early part of the twentieth century, there were two wings of the conservation movement: the more pragmatic wing, symbolized by Gifford Pinchot, a close adviser to Teddy Roosevelt, and the more idealistic wing, the "preservationists," represented by Sierra Club founder John Muir.[9] Pinchot became the chief of the U.S. Forestry

Service and cofounder of the Yale School of Forestry. He held that private interests could utilize forests, in exchange for a fee, so long as certain conditions were met. In his book *The Fight for Conservation,* he wrote:

> The first great fact about conservation is that it stands for development. There has been a fundamental misconception that conservation means nothing but the husbanding of resources for future generations. There could be no more serious mistake. Conservation does mean provision for the future, but it means also and first of all the recognition of the right of the present generation to the fullest necessary use of all the resources with which this country is so abundantly blessed. Conservation demands the welfare of this generation first, and afterward the welfare of the generations to follow.[10]

Muir, a passionate writer, held that nature was divine and must be preserved for its spiritual value rather than developed.[11] The two men were friends until 1897, when Pinchot's stance on development and the commercialization of nature received public attention.[12] At that point, Muir chose to distinguish himself as part of an oppositional camp of "preservationists." An excerpt from Muir's 1911 memoir *My First Summer in the Sierra* showcases his rapturous concept of the land:

> We are now in the mountains and they are in us, kindling enthusiasm, making every nerve quiver, filling every pore and cell of us. Our flesh-and-bone tabernacle seems transparent as glass to the beauty about us, as if truly an inseparable part of it, thrilling with the air and trees, streams and rocks, in the waves of sun—a part of all nature.[13]

Teddy Roosevelt's administration benefited from both men's efforts.[14] Bolstered by the surveys, reports, and the public-relations impacts of Muir and Pinchot, among others, the Roosevelt administration set aside an unprecedented 42 million acres of national forests, 53 national wildlife refuges, and 18 areas of "special interest," including the Grand Canyon, for a total of 194 million acres of preserved natural resources.[15] Among them were the so-called midnight forests, 16 million acres in the West that Roosevelt and Pinchot slyly and heroically signed into protection as national forests in the final hours before an amendment to the agriculture bill made the creation of additional forest reserves illegal.[16]

No one disputes the fact that these men—and yes, it's only *men* who are mentioned in the annals of history, although surely there were female voices too—accomplished worthy gains for the environment. Despite Muir's "preservationist" dissent, the efforts of that era have gone down in history as the launch of the "conservation" movement.

But Muir's more radical legacy was revived some half century later in the figure of David Brower. Also of European descent, Brower tried to move defense of the environment back into the realm of morality and spirituality. During the fight to keep the Grand Canyon from being dammed, Brower collaborated with Freeman, Mander & Gossage, Jerry Mander's innovative social-issue advertising agency, to create the infamous headline: "Should we also flood the Sistine Chapel so tourists can get nearer the ceiling?"[17]

In his book *Let the Mountains Talk, Let the Rivers Run*, Brower advocated for "CPR" for the Earth: conservation, preservation, and restoration. He included this prescient passage:

Restoration means putting the Earth's life support systems back in working order: rivers, forests, wetlands, deserts, soil, and endangered species, too. . . . Human systems also need restoration. Let's rehabilitate the South Bronx, and all the other places like it across the Earth. To accomplish that, we must give the unemployed and the never-employed a stake in the wider restoration process. Let's also put environmental conscience into world trade and into our corporate thinking.[18]

Brower's generation of conservationists won the designation of Redwood National Park and Point Reyes National Seashore in California; it also got the 1964 Wilderness Act passed, which established areas in which humans could not permanently reside.[19]

However, it is neither Roosevelt nor Pinchot, nor even the eloquent purists Muir or Brower, who should be celebrated as the first or truest champions of American conservation and preservation. The fact remains that all of them owed an incalculable debt to the physical and philosophical legacy of indigenous peoples.

Unfortunately, that debt continues to go largely unpaid—and even unacknowledged—by the conservation movement as a whole. It's really a shame. Imagine the good that could be done and the healing that could occur if major conservationist organizations were to fully honor the contributions of this continent's original stewards. For example, the aforementioned 1964 Wilderness Act defined "wilderness" as places absent of humans, where "man is a visitor who does not remain." This language was well-intentioned, but not all of its consequences were benign. In the words of scholar Carolyn Merchant:

> As environmental historians have pointed out, this characterization reads Native Americans out of the wilderness and out of the homelands they had managed for centuries with fire, gathering, and hunting. By the late nineteenth century, following the move to eliminate Native Americans and their food supplies, Indians were moved to reservations. National parks and wilderness areas were set aside for the benefit of white

American tourists. By redefining wilderness as the polar oppo-site of civilization, wilderness in its ideal form could be viewed as free of people, while civilization by contrast was filled with people. Yet this was a far different view of Indians than had been the case for most of American history, where Indian pres-ence in the landscape was prominent.[20]

Imagine if the 1964 act had carved out an exception to allow Native Americans greater access to lands the government deemed "wild." What if it had acknowledged the fact that traditional Indian lifestyles would not undermine the ecosystems, but enhance them? Unfortunately, the act—as monumental as it is—failed to make adequate provisions for the needs or the wisdom of America's indig-enous people. In other words, one of the biggest conservationist vic-tories overlooked the needs of the continent's original, indigenous conservationists.

Even today, in an era when large conservation groups have count-less members, hundreds of millions of dollars, and scores of profes-sional lobbyists, their Web sites and publications do not generally foreground the connection between the beauty of America's land and the wisdom and rights of its original inhabitants. Thus, when Native Americans fight poverty, hostile federal bureaucracies, and the impact of broken treaties, those massive environmental groups are too often absent. Or silent.

From that perspective, Indian-killing Teddy Roosevelt may have set the enduring pattern for the racial politics of the conservation move-ment. Viewed in the harshest possible light, perhaps his goals could be summed up simply as: "Let's preserve the land we stole, but get rid of the peoples from whom we stole it." Sadly, many of his own words and actions indicate that he had this kind of attitude. Such are the limita-tions and blind spots of even the greatest of human heroes.

And yet it must be said in closing, were it not for the heroic efforts of the preservationists and conservationists, much of the remaining natural wonder of North America would already be paved over. Many of the distinctive plants and beautiful animals that define this country and continent would be nothing but photos in a history book. Generations to come will sing the praises of the conservationists—and rightly so—for standing up to the buzz saws and bulldozers, for protecting and defending "America the beautiful."

But history also shows that, for all of those invaluable contributions, environmentalism's early record is marred by a failure to honor the full humanity and contributions of this continent's original stewards. The conservationists stood up for the most vulnerable places—but not always for the most vulnerable people. And decades later, we will see a similar shortcoming in environmentalism's second wave.

THE SECOND WAVE OF ENVIRONMENTALISM: REGULATION

Although active at around the same time as David Brower and also hailed as a conservationist, Rachel Carson was the pivotal figure in launching the next wave of environmentalism in the United States. I call this phase the "regulation" wave. Like the conservation wave that preceded it, this wave accomplished much, but it too stumbled over issues of race, class, power, and inclusion in ways that have much to teach us.

Carson was different from those who came before her. She was a scientist, a marine biologist, and only the second woman to ever be employed by the U.S. Bureau of Fisheries. Although Carson, who came from a white farming family that owned sixty-five acres in Pennsylvania, was able to attend college, she had to help support her family throughout her life.[21]

Like Brower, she maximized the written word in the form of articles and books; she also broadcast, through radio, film, and television, her messages about threats to the environment and the health of humans and other critters. Her most famous contribution was *Silent Spring*, a revolutionary book questioning the advances of the chemical industry and technology—especially pesticides:

These sprays, dusts, and aerosols are now applied almost universally to farms, gardens, forests, and homes—nonselective chemicals that have the power to kill every insect, the "good" and the "bad," to still the song of birds and the leaping of fish in the streams, to coat the leaves with a deadly film, and to linger on in the soil—all this though the intended target may be only a few weeds or insects. Can anyone believe it is possible to lay down such a barrage of poisons on the surface of the earth without making it unfit for all life?[22]

Her poetic protest helped lead to the eventual domestic ban of the particularly nasty pesticide DDT in 1972, although she and her legacy have had to endure a slew of distorted, personal attacks ever since from chemical-industry giants including Dupont, Velsicol, and American Cyanamid. For example, Carson specifically addressed the use of DDT in combating malaria, although she did not urge a total ban; instead, she insisted that the chemical be used sparingly. Her opponents continued to attack her legacy into the 1990s (Carson died in 1964 of a heart attack, following her battles with breast cancer), blaming her for deaths caused by malaria after DDT was banned. Despite these smears, Carson's standing as the initiator of the modern environmental movement remains intact and undiminished.[23]

By 1962, the year in which *Silent Spring* was published, Americans were palpably experiencing the environmental impacts of the ramped-up industrialization that followed World War II, especially from disposable packaging, the planned obsolescence of an ever increasing number of consumer goods, and the subsequent explosion of waste. As just one of many examples, the Cuyahoga River in Ohio, which travels past the industrial cities of Akron and Cleveland, regularly caught fire because of the amount of floating debris and oil on the water. A June 1969 fire on the river, though, received national attention and combined with *Silent Spring* in ratcheting up public awareness about the interconnectedness of human industry and the natural environment.[24]

The very first Earth Day, in April 1970, brought unity to the diverse grassroots efforts that had begun fighting against polluting factories and power plants, raw sewage and toxic dumps, pesticides, freeway construction, oil spills, the loss of wilderness, and the extinction of wildlife. Campuses across the country rallied in the name of the environment and against polluters.

In 1970, in response to the public's concerns and persistent media coverage of disasters, President Nixon formed the Environmental Protection Agency (EPA)—by cribbing personnel from four other federal entities. From the Department of Health, Education, and Welfare came the heads of Air, Solid Waste, Radiological Health, Water Hygiene, and Pesticide Tolerance; from the Department of the Interior, Water Quality and Pesticide Label Review; from the Atomic Energy Commission and the Federal Radiation Council, Radiation Protection Standards; and from the Department of Agriculture, Pesticide Registration. Legislation passed in the regulation era included the Clean Air Act (1963) and the Air Quality Act (1967), Federal Water Pollution Control Amendments (1972), Environmental Impact Statements required by the National Environmental Policy Act (1970), and the Endangered Species Act (1973).[25]

Just four years after *Silent Spring* was published, another biologist named Paul Ehrlich came out with *The Population Bomb*. This book—and the zero population growth campaigns that followed it—also challenged the status quo in the United States. This time it wasn't industrial pollution, but human reproduction, that was under fire and linked to environmental degradation as well as sustainability, the availability of resources, and poverty. With allies inside the reproductive rights and women's rights communities, zero population growth helped legalize contraception, for both married and unmarried folks, as well as abortion, through 1973's landmark *Roe v. Wade* decision.[26]

And at the end of the 1970s, local parents discovered a toxic chemical dump buried under Love Canal in Niagara Falls, New York. Leaking underground waste was said to be causing myriad nervous disorders, miscarriages, and birth defects in more than half of the children born in the area between 1974 and 1978. The highly publicized tragedy led Congress to pass the Comprehensive Environmental Response, Compensation, and Liability Act, better known as the Superfund law. Superfund assessed a tax on petroleum and chemical industries to create a fund for cleaning up abandoned or uncontrolled hazardous waste sites.[27]

Yet despite the era's multitude of gains, particularly in the domain of legislation, this second wave of U.S. environmentalism had its disappointing aspects as well. And once again the culprits were race, class, and power. The movement to better regulate industrial society was, in its origins, almost entirely the purview of the affluent and white. As a result, it failed to see the toxic pollution that was concentrating in communities of poor and brown-skinned people, even after major environmental laws were passed. In fact, some people of color began to wonder if white polluters and white environmentalists were unconsciously collaborating. At times it seemed that both groups were willing to see the worst polluters and foulest dumps steered into black, Latino, Asian, and poor neighborhoods.

These unjust and unequal outcomes prompted activists representing people of color and low-income communities to speak out—forcefully. In the 1980s, a new movement was born to combat what its leaders called "environmental racism." Those leaders said, in essence: "Regulate pollution, yes—but do it with equity. Do it fairly. Don't make black, brown, and poor children bear a disproportionate burden of asthma and cancer. Regulate, yes—but do it justly." Their battle cry marked the beginning of a serious corrective movement insisting that the regulatory wave protect all people equally. And that corrective movement has now become a force in its own right; it is called the movement for environmental justice.

This aspect of the second wave has a proud history. It was born in 1987, when the Commission for Racial Justice of the United Church of Christ released a landmark report entitled *Toxic Wastes and Race.* Among the devastating findings was that three out of the five largest commercial hazardous waste landfills in the United States were located in predominantly black or Hispanic communities. These three landfills accounted for more than 40 percent of the estimated commercial landfill capacity in the nation. In addition, in communities with a commercial hazardous waste facility, the average minority percentage of the population was twice the average minority percentage of the population in communities without such facilities. Finally, three out of every five black and Hispanic Americans lived in communities with uncontrolled toxic waste sites, while approximately half of all Asian/Pacific Islanders and Indians lived with them.[28]

Racial justice activists also expressed real concerns about other elements of environmentalism's second wave. For instance, some pointed to the fact that zero population growth campaigns can sometimes have racist overtones and implications, particularly when they focus on trying to limit the number of babies born to Africans, Asians, and Latin Americans. Many Latinos and Asian Americans were offended by the fact that zero population growth's founder, Paul Ehrlich, was a longtime board member of the xenophobic anti-immigrant group Federation for American Immigration Reform.[29]

Even the Superfund victory turned out to be a point of pain for communities of color. The *National Law Journal* found that the amount of the penalty paid by the polluter for cleanup was an average of 500 percent higher at sites with majority white populations. Meanwhile, abandoned hazardous waste sites in minority neighborhoods took 20 percent longer to attend to than in white areas.[30] And because Superfund targets the new or current owner of a Superfund-listed site for damages, most businesses choose to develop on virgin, nonindustrialized land. Fearing massive liability, they tend to avoid buying one of the many abandoned industrial sites, known as brownfields, that linger in our inner cities. One result is even fewer job opportunities for the urban unemployed.

None of these consequences were intended, but stronger protections could have been built into these laws. The fact that much of the legislation went forward with minimal involvement from people of color practically ensured that racially imbalanced outcomes would result. And when disparities and negative unintended consequences started to surface, the original champions of change could have been more responsive and enthusiastic about modifying the laws they had passed.

The good news is that the environmental justice movement itself has made significant progress and inroads. It is led in a decentralized way by academics such as Robert Bullard;[31] African American activists like Peggy Sheppard and Beverly Wright;[32] Native American leaders like Winona LaDuke, Tom Goldtooth, and Evon Peter;[33] Latino leaders like Richard Moore and Alicia Marentes;[34] Asian American activists like Peggy Saika, Pamela Chiang, and Vivian Chang[35]—and too many more to possibly name. The community's very diversity has allowed it to make the connection between race and the environment, linking the movement to the civil rights movement, Native American struggles, and the labor movement and including such groups as tenant associations, farmworkers, and religious organizations. Unlike in the traditional environmental movement, churches have played a significant role. For instance, the United Church of Christ published its landmark report and also helped to organize the First National People of Color Environmental Leadership Summit.[36]

Environmental justice activists helped to expand the definition of the "environment" to include neighborhoods and the economic as well as social aspects of communities. They focused on "disproportionate impact": the relatively higher level of exposure to environmental hazards faced by the poor and people of color. They also advanced the concept of "environmental racism": how policies and regulations exclude people of color from decision making. Throughout the 1990s, the movement saw victories such as the ones in South Central Los Angeles against the LANCER, a trash-to-energy incinerator; in Louisiana against a uranium-enrichment facility and a proposed chemical plant; and in Kettleman City, California, against Chemical Waste Management.[37] Bill Clinton signed a 1994 executive order directing the government to address these unjust patterns.

Some members of the environmental justice community, like Carl Anthony of Urban Habitat, realized the need to reach out to bridge the chasm between "mainstream" (white) environmentalism and the opponents of environmental racism. Early in the development of the environmental justice movement, he said: "There must be a massive campaign at the national level explaining the benefits that [the majority white environmental groups] might have from addressing diversity and environmental justice. At the same time, there needs to be a campaign with community organizations to teach them why they should be concerned with the environment. We need to create more ability to see the interconnectedness of things."[38]

But Anthony's pleas went largely unheeded. As a result, since the 1980s, the United States has had a shameful secret: its environmental movement is almost explicitly segregated by race—the mainstream environmentalists are in one camp (mostly white) and the environmental justice activists in another (made up almost entirely by people of color). Without assigning blame to anyone on either side, it is safe to say that the entire second wave of environmentalism has been less powerful, less perceptive, and less transformative than it might have been, if the leaders on both sides had been able to overcome the divisions. Certainly, the major victories—like the Wilderness Preservation Act and the Superfund law—could have achieved more good for more people, had the initial proposals been grounded in the perspective of a broader, more cohesive movement.

The lesson from both the conservation and the regulation waves of the environmental movement is clear: unless everyone is included at the decision-making table, even the best-intentioned proposals miss chances to do good—and may unwittingly even do harm. Our challenge now is to build these hard-won lessons into the very DNA of the next wave of eco-activism. If we fail to do so, the next effort will probably come up short—and possibly doom us all. But if we succeed, the next wave of environmentalism will have the power to carry our whole society through a positive transformation like nothing we have seen in our lifetimes.

INVESTMENT AGENDA: THE THIRD TIME'S THE CHARM?

Given the history of racial apathy, exclusion, and even conflict, is there any reason to expect anything different from the latest upsurge of eco-activism? I believe there is. This new stage is grounded in the dynamics and logic of a big, new economic opportunity. The need to expand markets and secure the favor of government will give the new green leaders a tremendous incentive to pitch as broad a tent possible.

Already, the green wave is taking off as a financial locomotive. We're seeing consumers and investors flocking to carbon-cutting solutions like solar power, hybrid technology, biofuels, wind turbines, tidal power, fuel cells, green construction, and energy effeciency. Reporters and editors are moving their environmental stories from the back of the paper to the first page, above the fold. Corporations compete furiously with each other in showcasing their love of clear skies and lush forests. Venture capitalists are pouring billions into clean-tech and green-tech companies. Organic cuisine and natural products are flying off the shelves. And both the blue Democrats and the red Republicans are suddenly waving green banners.

In other words, solutions to the climate crisis are galloping from the margins of geek science to the epicenter of our politics, culture, and economics. Concern for the Earth and an embrace of ecological values are moving from the eco-freak fringe to the eco-chic mainstream. A sea shift is taking place in public consciousness and concern. As the new environmentalists move to the front of the line in public discourse, only two questions remain: Whom will they take with them? And whom will they leave behind?

This is the great moral challenge facing the movement for climate solutions—and the broader movement for ecological solutions as a whole. For instance, we know that climate activists eventually will convince Congress to adopt market-based solutions (like "cap-and-trade"). This approach may help big businesses do the right thing. But will those same activists use their growing clout to push Congress to better aid survivors of Hurricane Katrina? Black and impoverished victims of our biggest eco-disaster still lack housing and the means to rebuild. Will they find any champions in the rising environmental lobby?

We know that the climate activists will fight for subsidies and supports for the booming clean-energy and energy-conservation markets. But will they insist that these new industries be accessible beyond the eco-elite—creating jobs and wealth-building opportunities for low-income people and people of color? In other words, will the new environmental leaders fight for eco-equity in this new "green economy" they are birthing? Or will they try to take the easy way out—in effect, settling for some version of eco-apartheid?

Eco-Apartheid?

It is not too early to sound the alarm against the possibility of eco-apartheid. In that scenario, on one side of town there would be ecological "haves," enjoying access to healthy, morally upstanding green products and services. On the other side of town, ecological "have-nots" would be languishing in the smoke, fumes, toxic chemicals, and illnesses of the old pollution-based economy.

This kind of morally disgraceful, politically untenable, and ecologically unsustainable result is not far-fetched—at all. In fact, we can already see the early signs of it. As my colleague Majora Carter, award-winning founder of Sustainable South Bronx, explains it:

> The public image of the environmentalist is all about eating organic food, driving a Prius, and buying solar panels. And that's incredibly narrow and alienating. In the South Bronx and other poor neighborhoods, people don't have a sense of belonging to the environmentalist identity. It makes low-income communities of color say, "We can't do it, we can't afford it, it's something that we can never aspire to—nor do we necessarily want to." And that just won't work. Sustainable and green alternatives will really take off only as we reach economies of scale. And to do that, we need everyone's participation.[39]

To have everyone participating and benefiting equally—that's the alternative to eco-apartheid. That's what we call eco-equity: *equal protection and equal opportunity in an economy that respects the Earth.*

The sad racial history of environmental activism tends to discourage high hopes among racial justice activists. They doubt the new greens will become enthusiastic champions or reliable partners in pursuing an eco-equity agenda. Many working-class whites look at the green phenomenon and shake their heads; all they see are hippies or snobs. They too doubt whether all the green hoopla will make a difference in their lives.

However, more optimism is warranted. This new wave has the potential to be infinitely more expansive and inclusive than previous environmental upsurges. The reason for hope has to do with the very nature of the present wave: because it is centered on investment and solutions, it is a qualitatively different phenomenon. Although they will always remain important topics, discussions of race, class, and the environment today can go beyond how to atone for past hurts or distribute present harms.

Today we can also ask: How do we equitably carve up the benefits of a green future for our children? How do we expand the number of people who are moving from the old, gray economy into the new, green one—as workers, consumers, investors, and owners? The new eco-entrepreneurs need all the customers and supporters they can get. And that creates a different basis to engage in cross-race, cross-class dialogue.

That's why working people need to give the green wave a second look. The green economy is not just a place where affluent people can spend money. It is fast becoming a place where ordinary people can earn money. In fact, the only part of the U.S. economy that is growing—the only part of the economy that *can* grow, long-term— is the green part.[40] So the green wave's new products, services, and technologies could mean something important to struggling communities: the possibility of new green-collar jobs, a chance to improve community health, and opportunities to build wealth in a sustainable way. Besides gaining dignified and meaningful employment, ordinary people have the chance to become inventors, investors, owners, entrepreneurs, and employers in the new, greener world. Working people will have a powerful incentive to support a green-growth agenda as long as green partisans embrace broad opportunity and shared prosperity as key values.

And the new greens have every reason—and every need—to reach out and be inclusive. To illustrate this point, let me share with you a pictorial graph from presentations I have made to audiences across the country. I use this grid to shed some light on the race and class dimensions of the environmental movements of the past, present, and future.

The horizontal axis charts ecological progress from old, gray problems, on the left, to new green solutions, on the right. The vertical axis maps the human dimension, from poor and mostly people of color, on the bottom, to wealthy and mostly white, at the top.

In quadrant one (upper left), relatively affluent people worry about ice caps melting, rain forests disappearing, oceanfront properties vanishing, and polar bears drowning due to global warming. Their own basic needs are met. Therefore, they have the time to think more deeply about long-term, global ecological problems. They also have the means and capacity to defend the long-term interests of the defenseless species that can't vote or lobby for themselves. This quadrant corresponds to the mainstream environmental movement—in its conservationist and regulatory modes. (Of course, Native Americans traditionally share these concerns, and few of them are materially wealthy. I suppose it is their spiritual and cultural wealth that lets them focus on the big picture and the long term.)

In quadrant two (lower left), Americans of more modest means worry about local environmental problems: dirty air, polluted water, cancer clusters, childhood asthma rates, lack of access to fresh food, and the devastation that might follow a disaster like Hurricane Katrina. Their basic needs are not securely met. Therefore, they tend to focus more on the personal and immediate aspects of the ecological crisis. Even though they may not consider themselves environmentalists, they have the means and capacity to push for important environmental changes at the local level. It is in this quadrant that much of the environmental justice movement is located.

In the third quadrant (upper right), affluent people choose to abandon their SUVs and Hummers. Instead, they buy hybrid cars, solar panels, and other green technologies. This is the quadrant of business opportunities for wealthy people. It is also the quadrant of consumer choices for the affluent (who must pay a premium for the slightly pricier green goods and services). Activity in this quadrant has a double benefit. The wealthy shift their considerable resources away from economic activity that harms the planet—and toward economic activity that helps it. As early adopters, they also create markets and support start-ups—which over the long term bring down costs and let more people participate in the green economy.

In the fourth quadrant (lower right), working-class people are motivated to take on green-collar jobs and start green businesses. This is the quadrant of "work, wealth, and health" for people of more modest means. Here, former brownfields, depressed urban areas, and hard-hit rural towns blossom as eco-industrial parks, green enterprise zones, and eco-villages. Farmers' markets, community co-ops, and mobile markets get fresh, organic produce to the people who can't afford to shop at health-food stores.

All four of these quadrants are important and have tremendous value. One quadrant is no more important than any other in the big picture, but right now, in order to expand the coalition opposing global warming and grow the green economy to include millions more people, the time has come to focus energy on building the fourth quadrant. Once the green economy is no longer just a place for the affluent to spend money, once it becomes a place for ordinary people to earn and save money—nothing will stop it. And this country will meet—and more than meet—the dual challenge of growing the economy without hurting the Earth.

The Case for Eco-Equity

For global-warming activists, embracing eco-equity—and ensuring that as many people as possible concern themselves with the issues of the second and fourth quadrants—would be a politically brilliant move. In the short term, a more inclusive approach will prevent polluters from isolating and derailing the new movement. Opponents of change will actively recruit everyone whom this new movement ignores, offends, or excludes. To avoid getting outmaneuvered politically, green-economy proponents must actively pursue alliances with people of color, and they must include leaders, organizations, and messages that will resonate with the working class.

The real danger lies in the long term. The United States is the world's biggest polluter. To avoid eco-apocalypse, Congress will have to do more than pass a cap-and-trade bill. And Americans will have to do more than stick in better lightbulbs. To pull off this ecological U-turn, we will have to fundamentally restructure the U.S. economy. We will need to "green" whole cities. We will have to build thousands of wind farms, install tens of millions of solar panels, and retrofit millions of buildings. We will have to retire our car, truck, and bus fleets, which are based on combustion engines and oil, and replace them with plug-in hybrids and electric vehicles powered by a clean-energy grid.

Reversing global warming will require a World War II level of mobilization. It is the work of tens of millions, not hundreds of thousands. Such a shift will require massive support at the social, cultural, and political levels. And in an increasingly nonwhite nation, that means enlisting the passionate involvement of millions of so-called minorities—as consumers, inventors, entrepreneurs, investors, buzz marketers, voters, and workers.

Climate-change activists may be tempted to try to sidestep the issues of racial inclusion in the name of expedience—but eco-apartheid won't work. The green sector needs to break out of its elite niche and succeed on a broad scale economically. If the green economy remains a niche market, even a large one, then the excluded 80 percent will inevitably and perhaps unknowingly undo all the positive ecological impacts of the green 20 percent. And that excluded 80 percent will also likely vote down measures to boost green business at the expense of the rest of the economy. So eco-apartheid would represent a self-defeating cul de sac for the green movement; at best, it would be just a speed bump on the way to eco-apocalypse.

Any successful long-term strategy will require that the green wave fully and passionately embrace of the principle of eco-equity.

Now is the time for the green movement to reach out. By definition, a politics of investment is a politics of hope, optimism, and opportunity. The bright promise of the green economy could soon include, inspire, and energize people of all races and classes. And nowhere is the need for a politics of hope more profound than it is among America's urban and rural poor. More important, climate activists can open the door to a grand historic alliance—a political force with the power to bend history in a new direction.

To give the Earth and its peoples a fighting chance, we need a broad, populist alliance—one that includes every class under the sun and every color in the rainbow. By focusing on the fourth quadrant—and ensuring that as many people as possible have a financial stake in the green economy—we have a real shot at that outcome. The key is to ensure that, having learned the lessons of the past, a critical mass commits to ensuring that the green wave lifts *all* boats.

THREE

Eco-Equity

So how do we move forward when government regulations and subsidies are still weighted heavily in favor of the old, gray economy? The dead hand of "politics past" is blocking humanity's path to a livable economic future.

The U.S. government continues to give huge tax breaks and payouts to the oil, gas, and coal industries; meanwhile, the fledgling solar and wind industries are left to beg and plead—just to get extensions of their modest tax credits. Multinational corporations benefit from lopsided trade deals that protect capital and copyrights, but fail to protect workers and the environment. Thus our trade treaties and tax code reward economic titans who abandon their native soil to exploit peoples overseas desperate enough to accept starvation wages and toxic pollution. Sadly, most of the economic power we need to green the Earth is still in the hands of people with a "pillage and pave" mentality. And they have unleashed their lobbyists to further defend their prerogatives, extend their power, and prop up their positions. As a result, between the leather-bound covers of innumerable law books, our lawmakers have written—in ink—all of the assumptions of a suicidal status quo.

Therefore, unless people around the world change their governments' policies, even the best green entrepreneurs and coolest gee-whiz inventions will fall far short—*far* short—of their potential impact. Even legions of conscientious consumers, nonprofit do-gooders, and enlightened local officials won't be able to propel the world's eco-entrepreneurs to victory. At least, not on their own.

To truly take off, the heroes and she-roes of the green economy need the raw power of the public sector to clear the runway. They need government on their side—not on their backs (and certainly not on the other side of the battle lines). Until national policy stops rewarding the despoilers and the downsizers, no green enterprise or industry will be able to reach its full potential as a creator of U.S. jobs or a healer of ecosystems. And this fact should come as no surprise to anyone. After all, no major new set of modern industries—from the railroads, to nuclear power, to the Internet—has ever succeeded without government playing a powerful and supportive role.

Additionally, unless the government helps to steer jobs and investment in new directions, those who most need the benefits of a new, green economy are highly unlikely to get them. If the best of the green wave bypasses the most disadvantaged urban and rural communities, then low-income and marginalized places will miss out altogether on their one shot in this new century at a glorious rebirth.

And smart government action is necessary not only to ensure that the green wave lifts up the poor within the industrialized countries. To survive and thrive during this period of wrenching changes, the Global South also needs wealthy governments to immediately begin making it a priority to assist them. To best develop their own economies while cutting carbon and restoring the Earth, our sisters and brothers in the developing world need the full engagement and cooperation of Western governments.

Therefore, the transition to a green-collar economy is not only a matter of economics and entrepreneurship; it is also a matter of policies and politics. The green-collar economic revolution cannot succeed without a corresponding realignment in the public sector. In addition to the power of business innovators, consumers, and grassroots champions, the green economy must add the might of government.

Turning the world's governments green will not be an easy task. And it will be especially challenging in the United States, given the entrenched political power of the old polluters and the overwhelming "business as usual" inertia inside the D.C. Beltway. To create a pathway to a livable future, a mobilized U.S. citizenry will have to march into the halls of power and rewrite the rules—at every level of government.

We cannot be naïve about the obstacles. A people's movement strong enough to achieve that aim would have to quickly become as big, sophisticated, and morally appealing as the greatest democratic movements of the last century. And yet building just such a movement is the central challenge—and the highest calling—of our time.

Success in this world-historic endeavor will require genius, courage, a Herculean effort—and a great deal of luck too. But we must begin. Fortunately, we have good examples and role models to guide us along the way.

When facing grave dangers in the last century, our parents and grandparents routinely faced up to the peril, overcame cynicism, and beat the odds. Today, as new generations climb onto the world stage, we are blessed to be able to learn from the heroic examples of those in the past who faced down totalitarianism, beat back an economic depression, and ended overt racial apartheid and colonial oppression in most parts of the world.

Their proud histories teach us that a successful movement for change requires three things. First, change must be grounded firmly in moral *principles*. Second, change must move rapidly to reinvent and realign *politics*. And third, change must effectively pursue and implement smart *policies*. Those of us who are concerned about the future must take those lessons as our own instructions and guideposts as we go forward.

What can ordinary people do to support the transition to an inclusive, green-collar economy—not just as smarter consumers, but as fully engaged citizens, informed voters, and active community members?

PRINCIPLES

Any movement that seeks enduring, transformative change must be founded on enduring, transformative principles. For example, the American settlers who led the revolt against British rule announced early on their commitment to a timeless principle, "No taxation without representation." And in 1776, that simple ideal became the anvil upon which ordinary people shattered the shackles of colonialism.

The labor movement also committed itself to timeless ideals: fair wages for all, safe working conditions, the right to bargain collectively, and dignity for all workers. Supported by those four pillars, labor activists were able to invent or help create practically everything good in the world's market economies—from the "middle class" to the "weekend." Labor's struggle continues into the new millennium, guided by those same unalterable principles.

In the United States, the founders of the civil rights movement declared their steadfast commitment to the principle of equality between the races. The women's rights movement devoted itself to the ideal of social equality between the sexes. The movement to liberate lesbians, gays, bisexuals, and transgendered people from the pain of ostracism and persecution is founded on the simple conviction that all persons, regardless of their sexual orientation or gender identity, should be equally free to love and form families without fear. The movement to protect the rights of immigrants is rooted in another simple idea, that all workers and families should be afforded basic rights and respect, no matter where they were born, what color their skin is, or what language they speak.

It should be noted that the political left has no monopoly on social-change movements that are rooted in unwavering principles. For instance, conservative leaders like Margaret Thatcher, Ronald Reagan, and Newt Gingrich helped guide the New Right to political victory over "big state" liberalism in the 1980s and 1990s. Their core principles—upholding lower taxes, smaller government, a strong military, and "traditional" values—continue to be the touchstone ideals of the conservative movement today.

Nor is the industrialized world alone in generating social-change movements grounded in clear principles. For instance, Mohandas Gandhi's independence movement was based on a core belief in equality between the Indian and British peoples—and nations. South Africa's anti-apartheid movement, led by Nelson Mandela, was committed to the ideal of a "free South Africa"—democratic, nonracial, and nonsexist. The clarity of those movements' principles guided their actions over decades, helped to attract global support, and ensured a generally positive outcome in their countries.

History teaches us that it is impossible to guide a complex series of deep changes in culture, economics, and law without first grounding efforts in a set of unchanging ideals. Successful movement leaders often end up employing a dizzying array of tactics; sometimes they are forced to make zigzag shifts in their short-term aims and mid-term goals. Yet their bedrock principles do not change. Their strategies may be complex, but their ideals remain simple and clear.

The movement to create eco-equity in the world will continue, through many ups and downs, for decades. Social-uplift environmentalism will not triumph in one day. So it behooves us to take some time to clarify the principles upon which we must rest our efforts to create an inclusive, green economy: equal protection, equal opportunity, and reverence for all creation.

Principle 1: Equal Protection for All

As we move into this age of ecological challenge and opportunity, our first principle must be "Equal protection for all." This ideal is key because, in an ecological crisis, those individuals, families, and communities without money and status will always be hit first—and worst. When the floodwaters rise, fires rage, droughts parch, or superdiseases attack, the most marginal cannot afford to get out of harm's way. They cannot afford to protect themselves. And still worse, once the crisis has passed, they are least able to bounce back, to rebuild, to recover. Therefore, as dangers multiply, we must revive—as a cornerstone commitment in our national life—the deep principle of equal protection.

Yet for decades our society has been moving in the opposite direction. Starting in the 1980s, it became fashionable to pretend that every social problem could be solved at the individual level. U.S. president Ronald Reagan famously declared, "Government is not the solution; government is the problem."[1] Most voting citizens looked at the practical flaws, bureaucratic dysfunction, and moral quandaries inherent in the liberal welfare state—and found that, in their hearts, they agreed with him.

Soon thereafter, Democrats joined Republicans in sneering at the idea that government should play a strong role in sheltering and shielding the disadvantaged. They even backed away from programs—like funding world-class public schools and affordable college tuition—that helped the middle class. Many leaders from *both* major parties began telling the public that there were few government solutions or collective answers to our problems; what we really needed, they suggested, was a return to strong families and "rugged individualism."

So it was that, throughout the 1990s, the federal government rolled back its commitment to protecting the most vulnerable members of our national family—by undercutting organized labor with global trade deals that favored big corporations, backtracking on affirmative remedies for victims of racial discrimination, and ending the federal right to welfare assistance for low-income mothers. In 1995, Democratic president Bill Clinton himself put it bluntly: "The era of big government is over."[2]

However, all was not well in America. Problems continued to mount in our health-care system, public schools, natural environment, and job market. Homicide and drug abuse continued to cut short young lives in urban America; suicide and meth addiction did the same among rural and suburban youth. Those who could afford it crawled down bright, digital wormholes. In the 2000s, they tried to cover the emptiness in their lives with flat-screen TVs or plug the holes with iPod earbuds. But nothing worked.

Still, our politicians and pundits never flinched. They went on assuring the public that everything would work out just fine. We all just needed to be a little bit more "rugged" and a little bit more "individual." That's all.

And if it turned out that some of the people in our midst just couldn't cut it, well then the government had a moral obligation to simply let those people "sink or swim." In the end, exposure to the harsh discipline of unchecked market power would be good for their lazy, little souls. Yes, indeed, this nationwide "tough love" austerity program was the key to a brighter future for all, even for the poor and disadvantaged.

Those who doubted that this path was morally defensible got a prompt rebuke. The airwaves were soon flooded with "prosperity preachers," each giving God's own blessing to the new, hardscrabble arrangement. Megachurch pastors with megawhite teeth assured their far-flung flocks that, with the right amount of prayer and the right mental attitude, great abundance, tons of wealth, and high profits were sure to be enjoyed by all.

So we ordinary people decided to give it a try. And as problems piled up for the country (and difficulties accumulated in our own lives), we ran after every solo solution we could find. We worked longer hours. We worked extra jobs. We hocked our homes. We bought lottery tickets. We sought shelter under a house of credit cards. And yet our expenses and troubles kept on rising.

Nonetheless, the very idea of a renewed government role in fixing these problems—indeed, the very notion of broadly shared problem solving of any kind—seemed outlandish, outdated, out of the question. Thus when we saw our own relatives and neighbors struggling financially, we blamed them for lacking sufficient pluck and guile to succeed. But we were fair about it. When our own debt and health crises began to threaten our own lives and dreams, we even blamed ourselves.

Then one day, something horrible happened. And that tragedy exposed our folly for all the world to see. In late August 2005, we turned on our television sets—and were shocked to see an American city under water. The aftermath of Hurricane Katrina—which drowned New Orleans and much of the Gulf Coast—left thousands of people dead and tens of thousands homeless. The storm's damage was magnified by faulty levees, which collapsed. Overnight, floodwaters swept away an iconic global city.

It was hard to fathom. But then the truly unthinkable happened. As the survivors struggled to stay alive in the waterlogged ruins awaiting help—none came. Not on the first day. Nor the second. Nor the fourth. Nor the fifth. For days on end, we saw desperate, hungry, and frightened people—crowded into the Superdome, trapped on rooftops, holding babies, waving American flags. They lived in the richest country in the history of the world, yet somehow that nation was unable to deliver to them a drop of water for their tongues or a scrap of food for their children's mouths.

Over time, tragedy curdled into spectacle. The submersion of New Orleans became the world's sickest reality TV show: "Will this woman *drown*? Will these people *starve*? Will this infant *ever* get a bottle of formula? . . . How many of these poor black people are going to *survive* the aftermath of Hurricane *Katrina*? . . . Stay tuned to find out! We'll be right back—*after* these messages!"

The world stared in disbelief. The truth slowly sank in. The citizens of that once proud city had been left to the tender mercies of what could only be called a "free market" evacuation plan. Everyone who owned a functioning car (and who had a working credit card) was perfectly able to flee. But those who didn't own private vehicles, those who didn't have credit cards or savings accounts, those who were two days shy of a pay day that might have let them buy a full tank of gas were left to face the floodwaters, alone. No matter how the TV talking heads tried to spin it or explain it away, the reality was painfully simple: an awful flood had come, and the United States had left behind its poor, its black, its disabled, its infirm—to "sink or swim."

The catastrophe of New Orleans was not the result of a deliberate act of malice by any person or party. In many ways, it was, however, the logical, necessary, and inevitable outcome of the kind of politics that both major parties have been promoting for two decades. It was a concrete manifestation of a mentality that says that we are not, in fact, our sisters' and brothers' keepers. Years of neglect of the nation's infrastructure and inattention to the needs of the poor combined with the colossal distraction of a military occupation of Iraq all added up to one thing: a government too hollowed out to competently perform its basic functions in a crisis.

The revulsion that gripped the country was total. The Bush administration's approval numbers started heading south that week and never recovered. Fifteen months later, U.S. voters walked into voting booths and ended fourteen years of GOP control of Congress—and terminated six years of one-party Republican rule in D.C. Few politicians referenced Katrina directly; the country was too deeply ashamed to consciously dwell on the images from those nightmare days. But the public had had enough of Bush's smirking and shirking—his incompetence and ho-hum contempt for life. The truth is that George W. Bush's presidency drowned in the floodwaters of Katrina.

And yet the hard work of exterminating the overall mentality that led to the abandonment of Katrina's victims still remains. In a time of climate chaos and dwindling resources, our society will have to face many more moments of danger: superstorms, intensifying wildfires, droughts, shortages of food and water, rising sea levels, new pathways for the spread of disease. An energy crunch could lead to global economic turmoil, leaving millions drowning on dry land. In a post-Katrina world, we must remember those left behind in the Gulf Coast disaster and say, "Never again!"

To be sure, all individuals must take full responsibility for their own lives, strive to live up to their full potential, and do their fair share of the work. That much goes without saying in a sane and healthy society. But there are some dangers that are too big for any individual to overcome, especially the most vulnerable among us. So in an age of floods, we must reject any philosophy that would tempt us to tell people in wheelchairs to "sink or swim." We must embrace, instead, the principle that says: "We are all in this together—come what may." On that basis, we can truly honor the principle of equal protection for everyone.

Principle 2: Equal Opportunity for All

The task at hand is not just to win equal protection from the worst of global warming and the other negative effects that go hand in hand with ecological disaster. It is also to win equal opportunity and equal access to the bounty of the green economy, with its manifold positive opportunities.

Wonderful developments in our economy are underway: solar power, wind-generated energy, organic food, improved mass transit, high-performance buildings, and more. All of these developments will deliver benefits to the Earth and society as a whole. If the architects of the green economy honor the principle of equal opportunity, they can also deliver help and hope to those who most need new jobs, new investments, and new opportunities.

Our business and political leaders will launch tens of thousands of new green enterprises and initiatives. Each time they do, they must ask the question: How can we make this effort inclusive, ennobling, and empowering to people who were disrespected in the old economy? How can this effort be used to increase the work, wealth, health, dignity, and power of our society's disadvantaged?

It is not yet fashionable in eco-elite circles to pay much attention to issues of social justice and equity. The potential is there. Living mainly in Hollywood, Silicon Valley, the San Francisco Bay Area, Seattle, and Boston, many of the architects of the green economy have photographs of Mohandas Gandhi on their walls. They consider themselves tolerant and open-minded people. Almost all of them, if asked, would confess to a deep respect for Dr. Martin Luther King, Jr., and the civil rights movement.

So it may be worth pointing out a strange set of facts. Dr. King is a global hero because he marched and died to racially integrate the last century's economy—even though that economy was based on the old pollution- and poison-based technologies. He made the supreme sacrifice; he laid down his life to ensure that the old economy—flawed though it was—had a place for everyone.

And he was not alone. Those buses that the freedom riders risked death to integrate—they were not using biodiesel fuel or hybrid engines back then. Those lunch counters that the civil rights activists risked beatings and arrests to open up to everyone—they were not serving organic tofu. Those schoolhouses, which little black children risked pain and humiliation to integrate—they were not green buildings with solar panels on them. No, the civil rights champions all risked their lives to win equal access to an economy that—in retrospect—was undermining the health of the planet. Their calling and achievements were undeniably among the noblest in human history.

If the crusade to racially integrate the dirty, gray economy represented the height of nobility in the last century, then how morally compelling is the calling to build an inclusive, green economy in this one? If Dr. King and other activists were willing to face attack dogs and fire hoses and murderous mobs to get everyone included in the pollution-based economy, then what should you and I be willing to do today to ensure that the new, clean, and green economy has a place in it for everyone?

It is important that we wrestle with these questions consciously and openly—before the greening of the world's economies proceeds irretrievably along the same lines as the unjust, unequal, gray economy. There is no racist governor standing in the warehouse door, blocking solar-company CEOs from hiring urban youth. There are no white-hooded hoodlums insisting that health-food stores charge prices for healthy food that low-income parents could never afford. On the other hand, there is no Bull Connor preventing African Americans, Asians, Native Americans, Middle Easterners, or Latinos from joining the movement to reverse climate change.[3] The barriers separating us from each other are wafer thin—and largely of our own making.

In pursuit of equity, today's rainbow-colored generations will not have to break into a closed, pollution-based economy. We have the option, instead, of cocreating an inclusive, green economy together. No one can stop us from doing so—except ourselves. The fact is, if we do wind up with some version of eco-apartheid in the United States and the industrialized countries, it will be because good people who knew better simply failed to do better.

We cannot afford that kind of moral shortfall. To solve our global problems, we need to engage and unleash the genius of all people, at all levels of society. Some of the minds that can solve our toughest problems are undoubtedly trapped behind prison bars, stuck behind desks in schools without decent books, or isolated in rural communities. A green economy that is designed to pull them in—as skilled laborers, innovators, inventors, and owners—will be more dynamic, more robust, and better able to save the Earth.

It perhaps goes without saying that our first two principles—equal protection and equal opportunity—go hand in hand. Especially for the most vulnerable, we have a duty to do two things: we must minimize their pain *and* maximize their gain. We are one human family. So on a good day, we should not leave anyone out. And on a bad day, we should not leave anyone behind. We should not accept a world where people of color and low-income people are always first in line for everything bad and then are left to benefit last and least when it comes to anything good.

Everyone must be allowed to share equitably in the benefits and the burdens, the risks and the rewards, of our transition to a more survivable economic system. That ideal must undergird and accelerate our commitment to equal opportunity and equal access in the green-collar economy.

Principle 3: Reverence for All Creation

The traditional environmental movement has wisely impressed upon the public at large the value of nonhuman life and the natural world. It insists that we don't have any throwaway species or throwaway resources. Those of us who labor to build the green-collar economy should affirm that insight and echo that conviction. And we should take it one step farther.

It is true that we don't have any throwaway species or resources. We don't have any throwaway children, throwaway neighborhoods, or throwaway nations either. Therefore, the green economy must do more than reclaim thrown-away stuff. It must also reclaim thrown-away lives and thrown-away places. And it must reclaim the thrown-away values that insist we are all members of one human family, with sacred obligations to each other.

In the United States, especially, we have strayed far from these truths. The following facts are worth repeating. We represent only 4 percent of the world's population, but we are responsible for 25 percent of the world's greenhouse gases.[4] And we now jail more than 25 percent of the world's prisoners.[5] In other words, one out of every four carbon molecules superheating the atmosphere has our name on it, and one out of every four people locked up anywhere in the world is locked up in a U.S. jail or prison. Some say that number is closer to 50 percent. This is a disturbing testament to a profound moral failing: we are functioning as if we have a disposable planet—and disposable people.

We know deep inside us that all beings have value. All people are precious. All of creation has sacred, inherent worth. We must take a stand in defense of the children of all species—including our own.

At some level, this stand is purely self-serving. Obviously, our desire to survive as a species dictates that we become much better stewards of—and partners with—the billions of nonhuman species with which we share this planet. The human family has invaluable friends and irreplaceable allies in the plant and animal worlds. We cannot continue indefinitely abusing those relationships. We cannot continue to tug at the web of life without tearing a hole in the very fabric of our earthly existence—and eventually falling through that hole ourselves.

At the same time, let's be clear. Creation is not to be revered simply because it is useful to the human species. Our commitment must be deeper than a desire merely to maximize the utility of other living beings and ecosystems for our own desires and pursuits. Creation has an independent value beyond and irrespective of us.

For example, imagine that someday our intergalactic viewfinders are able to reveal to us another planet as rich, thriving, and gorgeous as ours. The entire human family would be awestruck. We would stare at the images, endlessly variable and vibrant. We would love the beauty of that planet, wonder at its mysteries, treasure its manifold species. We would name our children, our buildings, and even our sports teams after its wonders. And even if no human ever were able to set foot on that faraway place, even if no corporation were able to extract its resources, even if no intrepid explorer could jump into a rocket ship and open up a McDonald's restaurant there—all of humanity would see the intrinsic worth and celebrate the miracle of such a planet's very existence.

Now imagine a situation in which that distant gem were somehow imperiled—say, by a massive asteroid whose trajectory was sure to destroy it. All of humanity would experience anxiety and foreboding. And even though no human had ever touched its trees' leaves, walked its shores, or cradled its furry offspring, such a planet's sudden obliteration would move the whole world to tears.

Let's take it a step farther. Imagine—on that day of global shock and of mourning—that someone standing at the water cooler in your place of work were to shrug and say, "Well, you know, that planet didn't add anything to Earth's GNP, so what was it really worth to us, anyhow?"

Can you imagine the reaction? Such a person would be exposed, in the eyes of everyone, as a fool—and a dangerous kind of a fool at that. Something deep within us recognizes that the true worth of creation can never be reduced only to its value in the human marketplace. And yet it is precisely the people who think like that watercooler cretin—those who try to reduce every value to a dollar value, those who try to measure beauty with a calculator—to whom we have given great authority in our national life and global affairs.

We need not deny the economic value of the Earth's resources. And thankfully, there are environmentalists who express their love for the Earth by quantifying in monetary terms nature's "services" (the dollar value of bees pollinating plants, of trees filtering water, etc.). In the public debate, they are doing a good thing. But in the quiet of our own souls, let us never forget that the full beauty, value, and the mystery of creation can never be captured on a spreadsheet.

We need a much deeper understanding of exactly what it is that our industrial society, in its present creation, is jeopardizing. We need a more profound perception of what is at stake. Perhaps it is time to humbly confess the wisdom of the Native Americans, from whom our forebears stole these lands. The original Americans knew and tried to teach their conquerors: "We don't inherit the Earth from our parents; we borrow it from our children. The Earth doesn't belong to us; we belong to the Earth." Reclaiming and reaffirming their wisdom—which is truly the wisdom of all indigenous people everywhere—is likely the first step toward discovering a survivable future for our children and grandchildren.

Those of us in the West have tended to limit our reverential awe strictly to the lands surrounding Jerusalem, the crossroads of the great monotheisms. Anyone who has visited that region can attest to its majesty and spiritual power; the world can rightfully call that place the Holy Land. And yet, from another perspective, all land is holy land; every people is a chosen people, divinely loved; and every creature—no matter how humble—is a signature work of the Creator's own genius. Recognition of those facts is the key to showing proper reverence for all creation—and encouraging all humans to tread more lightly and respectfully on what my grandmother always called "God's green Earth."

So those are the three pillars of the new "social-uplift environmentalism": equal protection for all people, equal opportunity for all people, and reverence for all creation. These are the lasting principles upon which we can build a modern movement to birth an inclusive, green-collar economy—here and around the world.

FOUR

The Green New Deal

T O BIRTH A just and green economy, our society needs the government to act as an effective midwife. And to get the public sector to play that role, champions of the green economy need a powerful political movement—one that is grounded in the kind of principles discussed in the last chapter.

Yet principles alone do not generate successful movements. Movements for political and social change also need a strategy— with long-term goals, enduring coalitions, and an effective mode of operating. (They also need a concrete policy agenda, which we will lay out in Chapter 7.)

In this case, the transition to an inclusive, green economy must be supported by a political movement that aims to create a "Green New Deal" in the United States and other industrialized nations; forges a "Green Growth Alliance" to unite the best of business, labor, social justice advocates, youth, people of faith, and environ-mentalists (while paying special attention to the challenges of work-ing across old divisions of race and class); and advances a positive, solution-oriented "politics of hope."

GOVERNMENT AS PARTNER

The time has come to reimagine and re-create the New Deal. The last time a serious economic crisis gripped the country was during the Great Depression. Early in that crisis, President Franklin D. Roosevelt took office and ended a generation of Darwinist social policy—in his first hundred days. With the support of a broad co-alition, FDR used the government's power to help the people, stimulate the economy, and restore the environment. His so-called New Deal represented a new arrangement in society with a more balanced division of authority between government, business, and civil society. Up until the mid 1930s, unregulated financial interests had been running amok—to the detriment of all (including, ultimately, much of the business community itself).[1]

Economists still debate the ultimate effectiveness of the New Deal's many programs. But for those who received a warm meal, signed up for Social Security, had their spirits lifted by murals and theater, or formed lifelong friendships while creating our national parks as Civilian Conservation Corps (CCC) members—there was not much debate. There was only gratitude for the sense of solidarity and common purpose in the face of national calamity—and a renewed sense of confidence in the future.

Today, we are entering a new period of national and global challenge; already, our society is being impacted by ecological, social, spiritual, and economic crises. To resolve them, the federal government must act boldly and comprehensively. A temporary tax credit here or there, briefly benefiting one or another clean-energy industry, is not enough to deal with the energy crisis. And a patchwork of job-training programs haphazardly assembled and rarely aligned with actual job opportunities is not going to move the needle on the jobs crisis.

We need an entire suite of programs—intelligently coordinated. We need a complete set of policies and programs that would accelerate a market-led transition to a cleaner, greener, and more just economy—creating jobs, renewing hope, and strengthening community in the process. In other words, the time has come for a "new" New Deal. And this time it should be a green one.

This, however, is not the only difference that we can imagine, as we fashion the New Deal 2.0. This time, we can also imagine a much wiser, smarter role for government overall. After all, despite its positive features, no one can deny the shortcomings of the last century's "welfare state," which the New Deal helped create in the United States. At the same time, no one can argue that this century's "warfare state" has been a vast improvement either. We need a new model, a new role for government in helping society to meet its defining and fundamental challenges.

For too long, political debate has been stale because it has been premised on a false choice. The left—in effect—has argued for big, clunky, compassionate government à la that of Lyndon Baines Johnson. The right—in effect—has argued for big, clunky, warmongering government à la that of George W. Bush. But most of us do not want government as a nanny. Nor do we want the government as a big RoboCop bully. We do not want the government to create a new bureaucracy to fix every problem. We are happy to place our faith in the power of ordinary people to do extraordinary things. We just want government to be a smart, supportive, reliable partner to the forces that are working for good in this country.

We know that society is going to have to meet some huge challenges in the coming period. The individuals, entrepreneurs, and community leaders who will step up to make the repairs and changes are going to need help. As they strive to meet world-class challenges, they will require and deserve a world-class partner in our government.

And that central insight—the idea of "government as partner" to the innovators, the scientists, the eco-entrepreneurs, the neighborhood heroes, the ones who are close to both the problems and the solutions—is the key to understanding how a Green New Deal might function. It would be born out of the knowledge that government can't do everything, but that government can play a key role as a partner to those who are trying to do the right things—namely, the entrepreneurs and community leaders who trying to solve the problems we face.

Government can become a much better partner to the eco-entrepreneurs who are trying to bring world-saving innovations to market by giving them permanent and reliable tax breaks, putting exponentially more research money on the table, making polluters pay for carbon emissions, and providing green employers with a well-trained, green-collar workforce. Government can be a better partner to civic leaders and community groups trying to solve neighborhood problems by helping to finance money-saving weatherization and solarization for low-income homes, reinvesting in science and math programs in public schools, supporting vocational and technical training in the green trades, and shifting money from the failed incarceration industry to smarter, cheaper programs that get better results by focusing on emotional healing, economic opportunity, and rehabilitation.

The time has come for a public–private community partnership to fix this country and put it back to work. In the framework of a Green New Deal, the government would become a powerful partner to the problem solvers of the world—and not the problem makers. Were government to play such a role, it would represent a dramatic turnaround. That's because right now the public sector gives most of its love, respect, and money to the problem makers in our economy: the war makers, polluters, and incarcerators. They all get billions and billions of dollars in tax breaks and direct subsidies, while the renewable sectors—the job creators of the future—still get pennies.

And yet our problems keep getting worse, not better. The wars in Iraq and Afghanistan have cost nearly a trillion dollars—and the government says we are still not much safer. The economic gain from those industries that pollute the air and savage the land will be more than erased by the costly consequences of our destroying our ecological life-support systems and superheating the atmosphere. And evidence is beginning to show that the prison industry (which has more than doubled in size and cost over the past twenty years) is actually making neighborhoods less safe—both by failing to rehabilitate the people it locks up and by diverting money from the community-based programs that could.

For too long, the government has been a partner to the problem makers—and all we have to show for it is more problems. By advancing the idea of "government as a partner to the problem solvers," we can break through some of the stale debates and false dilemmas of the last century, and we can finally move our society from the present raw deal to a Green New Deal.

THE NEW COALITION

We cannot achieve the goal of a Green New Deal just by wishing for it. There is an existing power structure—call it the "military-petroleum complex"—that holds sway over our national economic, energy, and foreign policies. It is unlikely that the present high lords of oil, coal, and armaments will reverse course or give up their power without a struggle. A new force must emerge to realign American politics, transform the political landscape, and supplant the Texas/Pentagon axis.

Therefore "step one" in getting the government to support an inclusive, green economy is to build a durable political coalition— one that aspires, ultimately, to govern. Again, the New Deal period offers an important example. It was the broad, electoral, pro–New Deal coalition that moved the government onto the side of ordinary people, not FDR alone. Farmers, workers, ethnic minorities, students, intellectuals, progressive bankers, and forward-thinking business leaders all joined forces at the ballot box to support FDR and his congressional backers as they worked to revive the economy.

To accomplish our tasks today, we need a similar force: an electoral "New Deal coalition" for our time. Let's call it the "Green Growth Alliance." A Green Growth Alliance would be a broad, coalitional effort—fusing wise, compassionate forces in civil society with the enlightened self-interest of the rising green business community.

Its aim would be to put the government on the side of the people and the planet. The goal would be straightforward: to win government policy that promotes the interests of green capital and green technology over the interests of gray capital (extractive industries, fossil-fuel companies) in a way that spreads the benefits as widely as possible. The idea would be to resolve the economic, ecological, and social crises on terms that maximally favor both green capital and ordinary people.

Some in the environmental and social justice worlds may wince at our explicit inclusion—and even prioritization—of the needs and interests of green businesses. Many activists of all stripes are suspicious of any corporation; they have become hostile to the entire business community without distinction or exception. They say, "Big, greedy businesses led us into this global mess; we can't trust them to lead us out." They eye with deep suspicion a lot of the green advertising being paid for by companies that have historically been big polluters.

Much of their concern is understandable. Abuses of power by many big corporations—both directly against workers and the environment and indirectly through what amounts to legalized bribery in the political system—continue to have profoundly negative consequences.

But there is another side to the business community, rarely seen or celebrated. In recent years, organizations like the Social Venture Network, Business Alliance for Local Living Economies, Co-Op America, Green Business Alliance, Ceres, and the Investors' Circle have been gaining members and momentum. These are groups of financiers, investors, entrepreneurs, and business leaders who are committed—even in advance of any legislation or comprehensive federal support—to conducting their business in a manner that better respects both people and the environment. They have been inspired by visionary entrepreneurs like Paul Hawken and Ray Anderson. Joel Makower's GreenBiz.com has been cheerleading for them and chronicling their efforts for years.

As for how the government will separate the truly green companies from the pretenders—fortunately, the most trustworthy and socially conscious business leaders (not to mention investors, who don't want to be hoodwinked) are already far along in the process of defining tough criteria themselves. We can use their initial standards and definitions as starting points—and springboards.

The numbers are small right now, but the emergence of true "triple-bottom-line" businesses (which balance "profit, planet, and people" in their operations) is a potentially significant development. And it is just beginning. "Green" MBA programs like the Bainbridge Institute and Presidio School of Management are cranking out young business leaders with a different view about how to make money while making a difference. At the same time, increasing numbers of minority-owned enterprises (MBEs) and women-owned enterprises (WBEs) are expressing interest in "going green" and becoming part of the eco-revolution.[2]

These businesses—plus renewable-energy companies, firms selling conservation services, community-based cooperatives, nonprofit social enterprises, organic food companies, recyclers, and others—constitute business sectors with whom people of conscience can and should cooperate. Ordinary citizens and community members should actively help them win maximum support from the government—so that they can displace the despoilers and replace the polluters. And the faster we change the rules to aid the emerging sectors, the faster the old dinosaur companies will retool and get on board.

There will surely be an important role for nonprofit, voluntary, cooperative, and community-based solutions. But the reality is that we are entering an era during which our very survival will demand invention and innovation on a scale never before seen in the history of human civilization. Only the business community has the requisite skills, experience, and capital to meet that need. On that score, neither government nor the nonprofit and voluntary sectors can compete, not even remotely.

So in the end, our success and survival as a species are largely and directly tied to the new eco-entrepreneurs—and the success and survival of their enterprises. Since almost all of the needed eco-technologies are likely to come from the private sector, civic leaders and voters should do all that can be done to help green business leaders succeed. That means, in large part, electing leaders who will pass bills to aid them. We cannot realistically proceed without a strong alliance between the best of the business world—and everyone else.

That said, again, no business should be considered "green" just because it says it is. We need strong standards and clear criteria to weed out those companies that will seek out support while merely "green-washing" their same old bad practices. And society's support should not be unconditional. All legislation to boost green industry should also be strong on labor rights and civil rights. The captains of green enterprises should go beyond the letter of any law, enthusiastically seeking out the full diversity of the country in their hiring, promotions, and contracting. And, without undermining their need to stay profitable, they also should seek to locate their operations in places that need new infusions of jobs and capital. Only companies that work to meet those tough standards should be considered truly "triple bottom line."

On the civil society side, five main partners should make up the Green Growth Alliance:

1. *Labor.* Organized labor has been in steep decline over the past few decades, but it remains the best and most stalwart defender of working people's interests—in the workplace and beyond. Policies that lead to the retrofitting and rebuilding of the nation will give unions a tremendous opportunity to both expand and diversify their ranks. If the unions and green business leaders can identify win-win compromises on wages and other issues, they can work together to pass legislation that will help both sides.

2. *Social justice activists.* Legions of people have committed themselves to broadly shared opportunity for those who were left out of the old economy. They should be on the front lines working to create the new one. Advocates for economic justice, civil rights, immigrant rights, women's rights, disability rights, lesbian/gay/bisexual/transgendered rights, veterans' rights, and other causes should seize the opportunity to ensure that the new, green economy has the principles of diversity and inclusion baked in from the beginning.

3. *Environmentalists.* With their large organizations, broad networks, Beltway savvy, and large budgets, the mainstream environmental organizations have tremendous assets to bring to bear in the effort to green the country. They have a chance

to turn the page on decades of perceived elitism by working as better collaborators with other sectors of society. An exchange of knowledge, experience, and even personnel between the mainstream environmentalists and social justice groups would be healthy and invigorating for everyone.

4. *Students.* Students' energy and enthusiasm have already turned up the heat in the movement to prevent catastrophic climate change. Just a year ago, it was considered outlandish for anyone to call for an aggressive target like an 80 percent reduction in carbon emissions by the year 2050. But student-centered efforts like Step It Up, Focus the Nation, and the Energy Action Coalition[3] have already made "80 by 50" a mainstream demand—accepted by presidential candidates and even energy-company CEOs. As more racially diverse groups like the League of Young Voters, the Hip Hop Caucus, the Environmental Justice and Climate Change Initiative, and Young People For (YP4)[4] join the movement, the sky is the limit for the next generation's leadership role.

5. *Faith organizations.* The moral framework suggested by the three principles of social-uplift environmentalism (equal protection, equal opportunity, and reverence for all creation) should attract faith leaders and congregants. Many are looking for alternatives to some of the divisive fundamentalism that has taken up a great deal of airtime lately. The idea of "creation care" is a positive, alternative frame that can help faith communities move into action as a part of the Green Growth Alliance.

These five forces, in alliance with green business, can change the face of politics in this country.

The Importance of Including People of Faith

Just as some may recoil at the prospect of warmly embracing the business community, others may raise their eyebrows at the idea that religious organizations should play a strong and leading role in the greening of the country. Many environmental and social change activists say, "Well, I'm all for spirituality, but I reject religion." This is a perfectly reasonable and respectable personal choice. Yet it can often mask a deep resentment or even a hatred of the totality of organized worship—and a stereotyping of all religious people as stupid, dogmatic simpletons.

Too often those working for change quietly see religion itself as the enemy. They tend to reduce the great faiths of the world to their worst elements, constituents, and crimes—and then dismiss all other facts and features. Nothing pains me more than to hear so-called progressives snarl the word "Christian" as if it were an insult or the name of a disease. I grew up in the black churches of the rural South, listening to the civil rights stories of my elders. As children, we heard about the good and brave people who had poured their blood out on the ground so that we could be free. We learned how police officers had clubbed and jailed them. We learned how Klansmen had shot and lynched them. And how the G-men from Washington just stood by and doodled on their notepads.

We learned of marches and mayhem, freedom songs and funerals. We saw images of black women on their hands and knees, searching for their teeth on Mississippi sidewalks—crawling while still clutching their little American flags. We felt pity for the children who spent long nights in frigid jail cells, wearing clothing soaked by fire hoses, while their untended bones began to mend at odd angles. We saw pictures of black men like our fathers hanging by their necks, their faces twisted, their bodies rigid, their clothes burned off—and their skin too. And we saw photos of carefree killers sauntering home out of Alabama courtrooms—their white faces sneering and proud.

We learned how the very best of humanity had faced off with the very worst of humanity—each circling the other, under the same summer sun. Their epic struggle had elevated Southern backwaters onto the great world stage. And the fate of a people—and the destiny of a nation—hung in the balance, for all to see.

In the end, we cheered, for the righteous did prevail. Our parents and grandparents overcame—and then some. They performed one of the great miracles in human history; they transformed a U.S. apartheid into a fledgling democracy, tender and delicate and new. And today's social change activists proudly and eagerly celebrate the achievements of the civil rights movement. Rightfully so.

But one key fact seems to escape their notice. The champions of the civil rights struggle didn't come marching out of shopping centers in the South. Or libraries. Or high-school gymnasiums. They came marching out of churches, singing church songs. These people, these unimpeachable examples of audacity and accomplishment, were people of deep, deep religious faith. And when they prayed, it was through a long-dead Nazarene carpenter named Jesus Christ.

When progressives dismiss and disdain religious people, they are spitting directly in the faces of their greatest champions. This is why smug activists who treat the word "Christian" as a useful synonym for "dumb, mean bigot" do so much damage. They offend people of faith within our ranks. They needlessly cut themselves off from their own elders, families, and neighbors. And they deny the truth of how meaningful social change has most often come about in this country.

Worse, they leave powerful symbolism in the hands of dangerous practitioners of a less noble politics. It makes no sense to those seeking change to willingly surrender the language of the great faiths. In the end, it is nearly impossible to shrink our message of abundant love, hope, and faith into the tiny straitjacket of a sterile secularism anyway. Many activists are already turning, in the quiet of their own lives, to yoga and meditation, to self-help books, to alcohol and drug recovery programs that invoke a higher power, and even to the organized religions of their childhoods. Of course, no movement should force any particular brand of religious observance or spirituality on anyone else. But even secular activists sometimes seek for a power that is greater than our familiar "power to the people." And it should be okay to acknowledge that.

We do a great disservice to the cause of justice when we pretend that only the hateful represent the faithful. We can oppose theocracy without opposing all theology. We can denounce bias in the Christian church—whether the bigotry appears as sexism, homophobia, racism, or anti-Semitism—without pretending that all Christians are bigots. We can separate fundamentalism from faith in any of the great religions.

The United States is one of the most religious countries in the world. People of faith here have powered much of the social change in our nation's history—from the abolitionists up to the present peace movement. Imagine how powerful the Green Growth Alliance will be when it can claim as one of its pillars the millions of Christians, Jews, Muslims, and others who acknowledge and worship a justice-loving God.

Green Growth Alliance Seeds Already Sprouting

Fortunately, the Green Growth Alliance is not just a theoretical necessity. It is already becoming a practical reality. National organizations such as the Apollo Alliance and the Blue Green Alliance have already come on the scene, promoting good jobs in the clean-energy sector. The Apollo Alliance is an alliance of labor unions, environmental organizations, community-based groups, and businesses;

the Blue Green Alliance is a partnership between the Sierra Club and the United Steelworkers.

Former U.S. vice president Al Gore's Alliance for Climate Protection is also reaching out broadly to engage new sectors in the battle to avert catastrophic climate change. And the new kids on the block—1Sky and Green For All—are engaging important new constituencies like PTA moms and African American ministers. (I serve on the Apollo Alliance board, and I am a co-founder of both 1Sky and Green For All.) The bottom line is that the raw materials for a Green Growth Alliance already exist.

Of course, the very idea of "growth" itself will be challenged over time—whether it is green growth or any other kind. On a crowded planet, the very notion of economic growth itself—with automatic assumptions of increasing resource consumption and consumerism—is something that human society will someday be forced to abandon.

In the future, resource constraints and a growing population will force human society to adopt an even more sustainable model: a closed-loop, "steady state" economy premised on nearly 100 percent recycling of materials and 100 percent renewable energy. This economy will be designed to maximize well-being—not necessarily wealth. The growth we seek will be in steadily improving the quality of life, not steadily increasing the quantity of goods consumed.

However, at this stage, a quick leap to this kind of postgrowth, postconsumerist "eco-topia" is not possible. Let's take a lesson from history on this one point. The day after Rosa Parks refused to give up her bus seat, civil rights leaders could have demanded reparations for slavery, legalization of interracial marriages, and a massive redistribution of wealth. Such demands would have been justifiable—but foolish. "Maximum demands" like those would have created more resistance than support.

Thus early civil rights champions instead pressed "minimum demands"—for integrated buses, kindergartens, and lunch counters. The more militant Malcolm X attacked this approach as too timid. But the more modest demands knocked over the first dominoes, sparking mass movements that eventually tore segregation from U.S. law books, led to the War on Poverty, and helped stop a U.S. war.

Well, today, with Earth itself imperiled, we should be at least as savvy as our grandparents were. It's fine for the "Malcolm Xs" of the environmental and social movements to directly challenge the paradigm of growth and consumption—or to attack U.S. militarism. They are advancing important arguments. But to move large numbers, we also need smart, minimum demands or goals—like the calls for "Green-Collar Jobs for All," "Green Jobs, Not Jails," and "Greening the Ghetto First." Such slogans do not challenge everything that is wrong in the world. They simply point in the right direction. And they can inspire pragmatic, working people to take notice and get involved.

A family-friendly "eco-populism" can mobilize and unite millions who, at this point, would be turned off by a more extreme set of demands. Such appeals can get more people taking the first practical steps toward ecological sanity. The momentum will build, through those early efforts, for more comprehensive solutions. The later, bolder solutions will be even more effective, intelligent, and productive when they arise from the experience of millions of ordinary people who already have been actively engaged.

Meanwhile, the most urgent task remains: to rapidly shift society from our present suicidal, gray form of capitalism to a more viable and just eco-capitalism. Success will mean that the Earth will still be a livable place decades from now, giving those still alive the chance to pursue other improvements and transformations.

MORE ECO-POPULISM, LESS ECO-ELITISM

Although the movement for social-uplift environmentalism is on its way to forging an eventually powerful Green Growth Alliance, the very notion that a politics centered on green solutions could build a muscular, governing majority in the United States seems laughable at the moment. That is because the "green movement" itself seems to be the cushy home of such a thin and unrepresentative slice of the U.S. public.

When most people think of "green solutions," they are not thinking about a massive people's movement that can pick up the Capitol building in D.C., turn it upside down, and dump out all the legislators who are holding back a green economic renaissance. They are not thinking about the next best thing to a full-employment program, led by the private sector, that could put millions of Americans back to work retrofitting the country. They certainly are not thinking about the incredible pioneers—like Winona LaDuke, Majora Carter, and Omar Freilla[5]—who are daily bringing hope and opportunity to people of modest means.

Rather, they perhaps imagine a few Hollywood celebrities eating tofu, doing yoga, and driving hybrid cars. They envision affluent white people who care about nothing but polar bears and can afford to shop at health-food stores and put solar panels on their second home. Their minds leap to a high-priced market niche serving individual consumers who are willing to pay a premium for green goods and services so they can feel better about themselves and their many purchases.

Many of these caricatures are grossly unfair. Many lifestyle greens are actually just health-conscious, community-minded folks who earn middle-class wages. As for the rich ones, it is actually a good thing that wealthier people are now spending their dollars in ways that are healthier for the planet. They have helped to jump-start the market for the products and technologies that may help save the Earth.

Nonetheless, when many ordinary people hear the term "green" today, they still automatically think the message is probably for a fancy, elite set—and not for themselves. And as long as that remains true, the green movement will remain too anemic politically and too alien culturally to rescue the country. Enlightened, affluent people who embrace green values do a great deal of good for the country and the Earth—and they are making an importance difference every day. But nobody should make the mistake of believing that a small circle of highly educated, upper-income enviros can unite America and lead it all by themselves. Eco-elite politics can't even unite California.

If you doubt me, let's examine a recent statewide election in California to see how eco-elitism can actually set back environmental initiatives—even very thoughtful and well-financed ones, even in places where the overall support for environmentalism is relatively high. Everyone loves to praise GOP governor Arnold Schwarzenegger for signing global-warming legislation in 2007. Yet few discuss the fact that just a few months earlier, the majority of California voters rejected a clean-energy ballot measure called Proposition 87. That's right. Elected officialdom might be willing to dictate major green steps—or at least hold major green press conferences. But when Californians got the chance to speak up in the ballot booth in 2006, ordinary people said no.

This defeat holds many lessons for us, going forward. The idea for Prop 87 was brilliant in its simplicity: California would start taxing the oil and gas that oil companies extract from our soil and shores. This state-level oil tax would generate revenues of $225 to $485 million annually. And those dollars would go into a huge "clean-energy" research and technology fund—totaling $4 billion over ten years.[6] Many states and nations have similar extraction taxes.

California, however, would have been essentially alone in dedicating the revenues to inventing alternatives to carbon-based energy sources. Had the measure passed, California would have used money from oil to find a replacement for oil.

It was a brilliant idea. And at first, the measure was polling off the charts. Silicon Valley and Hollywood put $40 million on the table to ensure the measure passed. Al Gore and Bill Clinton campaigned for it. Victory was certain. But in the end, Californians voted the measure down 45 percent to 55 percent. Why? Mainly because big oil convinced ordinary Californians that the price tag would be too high for them to bear. The oil and gas industry spent $100 million warning that the tax would be passed along to consumers. They suggested that it would push gas and home-energy costs through the roof and hurt the poorest Californians.[7]

It was a predictable line of attack. It was also a false argument. Gas prices in California are not determined by oil extraction taxes in any one region or state; they are set mainly by the huge global energy market. Numerous other states and countries already have a similar tax, including Texas, Alaska, and Venezuela. One more teeny levy, in one state, in one country, would have had a minuscule or negligible impact on the overall world price of oil—or on California consumers.

And to the contrary, the benefits of a shift to cleaner energy would have helped the poorest in the state—significantly improving both their health and their chances for wealth. For one thing, disproportionate numbers of low-income people live near oil refineries and other sources of dirty-energy pollution. As a result, they suffer from higher rates of cancer, asthma, and other illnesses. Largely uninsured, they then pay through the nose for inferior medical care. In other words, the dirty-energy economy is literally killing poor people. A switch to cleaner energy could save untold lives.

Beyond that, a clean-energy economy actually is more labor-intensive—meaning, it creates more jobs. After all, somebody has to install and maintain all those solar panels, build all the wind farms, construct the wave farms, weatherize those millions of homes and office buildings. A green economy begins to replace some of the clunking and chugging of ugly machines with the wise effort of beautiful, skilled people. That means more jobs.

So there was a strong, eco-populist argument to be made for Prop 87. Switching to clean energy would have cost individual Californians little—but given working people better health and better jobs. Yet the campaign—led almost exclusively by well-intentioned do-gooders with few financial problems themselves—did not make these eco-populist arguments with any force. Instead, Prop 87 commercials yammered on about "energy independence"—which polling firms said was the best message. Maybe so. But in some demographics, people needed that message to be bolstered by a great deal of reassurance on the kitchen-table issues, and it never was.

Seeing the obvious opening, the polluters pounced. Big oil ran full-page ads in practically every African American newspaper in the state. The ads showed a black mom looking aghast at fuel prices while she tried to fill up her car. An NAACP official vocally opposed the measure, fearing economic damage to her constituency. And the scare tactics didn't alarm only black Californians. Across the state, the initially sky-high poll numbers for the initiative proved surprisingly fragile. The support for the measure was completely hollow. The Prop 87 proponents were not just outspent by the polluters; they were outmaneuvered.

And in the end, the biggest clean-energy ballot measure in the country went down to defeat—in California. That setback didn't just hurt the backers of Prop 87 or the Golden State. It set back the entire world's ability to invent new, clean-tech technologies and beat global warming.

The Eco-Elite Cannot Win by Itself

The defeat of Prop 87 should sound a clear warning for all of us as we work to birth a green, postcarbon economy. We all must recognize and celebrate the fact that well-off champions of the environment will be indispensable to any coalition effort. In fact, it is their business smarts, monetary resources, social standing, and political savvy that have propelled the green wave to this point. But at the same time, the eco-elite cannot win major change alone, not even in the Golden State. After all, if a Prop 87–style collapse is possible in California, what do you think will happen in the other forty-nine states?

To change our laws and culture, the green movement must attract and include the majority of all people, not just the majority of affluent people. The time has come to move beyond eco-elitism to eco-populism. Eco-populism would always foreground those green solutions that can improve ordinary people's standard of living— and decrease their cost of living.

The messaging must make it plain to the country that we envision a clean-energy future in which everyone has a place—and a stake. One way to do that is to speak to the economic and health opportunities that "ecological" solutions will also provide. Another way is to always show the many, many people of color and working-class Americans who are actively engaged in environmental struggles. If nearly every "green" initiative, TV program, or magazine cover excludes them, we are essentially handing millions and millions of people over to the polluters. We are essentially saying to big coal and big oil: "Please organize all of these people against everything green. Thanks!"

The nation has already passed a certain tipping point in eco-consciousness. But we should never underestimate the danger of rapid progress actually fueling a major backlash. Handled badly, green proposals can create the opposite of our much needed Green Growth Alliance. They can actually produce a backlash alliance— between polluters and poor people. The polluters are afraid of losing their immense profits and privileges. And low-income folks are just afraid of losing even more ground. Together they can derail the movement to green this country. We can, however, easily head that off, just by making sure that "green" includes all classes and colors.

NOT JUST HYBRID CARS—A HYBRID MOVEMENT

Bringing people of different races and classes and backgrounds together under a single banner is tougher than it sounds. The affluent have blind spots. The disadvantaged have sore spots. And both pose barriers to cooperation.

For instance, large and powerful constituencies of white, affluent, and college-educated progressives exist and are active in the United States. They are passionate about the environment, fair trade, economic justice, and global peace. Unfortunately, many do not yet work in concert with people of color in their own country to pursue this agenda; they champion "alternative economic development strategies" across the globe, but not across town. These people could be great allies in uplifting our inner cities, if they are given encouragement and a clear opportunity to do so.

On the other hand, the truth is that many groups of people of color do not want to work in coalition with majority white organizations and white leaders. Many fear betrayal; others resent chronic white arrogance. Cultural differences and power imbalances create tensions; some organizations are actually committed to a racially exclusivist ideology. Even though such organizations could benefit from additional allies and outside assistance, the very folks who could most benefit from a green opportunity agenda are loath to get involved.

Taken altogether, this means that the various U.S. social change movements today are still nearly as racially segregated as the rest of U.S. society. This is a moral tragedy. And it is a tremendous barrier to building sufficient power to move forward a positive social change agenda for anyone and everyone. Breaking through this standoff is a critical first step toward building a "New Deal coalition" for the new century—which would be the only thing dynamic, diverse, and powerful enough to overcome the shared obstacles to progress.

I have been trying to bridge this divide for nearly a decade. And I have learned a few things along the way. What I have found is that leaders from impoverished areas like Oakland, California, tended to focus on three areas: social justice, political solutions, and social change. They cared primarily about "the people." They focused their efforts on fixing schools, improving health care, defending civil rights, and reducing the prison population. Their studies centered on "social change" work like lobbying, campaigning, and protesting. They were wary of businesses; instead, they turned to the political system and government to help solve the problems of the community.

The leaders I met from affluent places like Marin County (just north of San Francisco), San Francisco, and Silicon Valley had what seemed to be the opposite approach. Their three focus areas were ecology, business solutions, and inner change. They were champions of the environment who cared primarily about "the planet." They worked to save the rain forests and important species like whales and polar bears. Also, they were usually dedicated to "inner change" work, including meditation and yoga. And they put a great deal of stress on making wise, Earth-honoring consumer choices. In fact, many were either green entrepreneurs or investors in eco-friendly businesses in the first place.

Every effort I made to get the two groups together initially was a disaster—sometimes ending in tears, anger, and slammed doors.

Trying to make sense of the differences, I wrote out three binaries on a napkin:

1. Ecology vs. Social Justice

2. Business Solutions (Entrepreneurship) vs. Political Solutions (Activism)

3. Spiritual/Inner Change vs. Social/Outer Change

The Marin County leaders tended to focus on the left side of the list; the Oakland leaders usually focused on the right side. And for some reason the people on both sides tended to think that their preferences precluded any serious embrace of the options presented on the opposite side of the ledger.

Increasingly, I saw the value and importance of both approaches. I thought to myself: What would we have if we replaced those "versus" symbols with "plus" signs? What if we built a movement at the intersection of the social justice and ecology movements, of entrepreneurship and activism, of inner change and social change? What if we didn't just have hybrid cars—what if we had a hybrid movement?

I came to believe that at the precise place where all these countercurrents converged was where we would find enough power to generate a Green Growth Alliance, displace the military-petroleum complex, and initiate a Green New Deal. But first I had to figure out how to engage African Americans, Latinos, and others from the urban environment who were resistant even to the idea of being a part of something calling itself "green."

I had two main breakthroughs in finding a way to move urban leaders. I call these breakthroughs "The *Amistad* Meets the *Titanic*" and "Crisis vs. Opportunity."

The *Amistad* Meets the *Titanic*

Most people who are committed to racial justice activism see themselves as rebels against racism. Perhaps they would most deeply resonate with Cinque, the hero of the slave-revolt movie *Amistad*. In that film, based on a true story, the righteous, enslaved Africans fight back and take over the slave ship.

The people at the bottom rise up and take over the ship—taking their destiny into their own hands. It's really a metaphor for the last century's version of racial politics. The slave ship is Earth, the white slavers are the world's oppressors, and the African captives are the world's oppressed. The point is for the oppressed to confront and defeat their oppressors. I took that as my mission.

But what if those rebel Africans, while still in chains, had looked out and noticed the name of their ship was not the *Amistad,* but the *Titanic*? How would that fact have impacted their mission? What would change if they knew the entire ship was imperiled, that everyone on it—the slavers and enslaved—could all die if the ship continued on its course, unchanged.

The rebels would suddenly have had a very different set of leadership challenges. They would have had the obligation not just to liberate the captives, but also to save the entire ship. In fact, the hero would be the one who found a way to save all life on board—including the slavers. And the urgency of freeing the captives would have been that much greater—because the smarts and the effort of everyone aboard the ship would have been needed to save everyone.

"*Amistad* meets the *Titanic*" has been an important bridge metaphor to help people who have been committed to an earlier model of racial justice activism understand their expanded leadership role and responsibilities—as the entire planetary ship is threatened with going down.

Crisis vs. Opportunity

A lot of environmental rhetoric remains rooted in "crisis" language. It is evident that people who already have a lot of opportunity are sometimes powerfully motivated to act by tales of a planetary crisis. But people who already live in a constant state of personal crisis are not so moved. In fact, they often have the opposite reaction to hearing about things like global warming. They will shrug, shake their heads, and say: "Well, it's just the end times, I guess. That means Jesus is coming back." And then they will change the subject or walk away.

But if you tell people who are living in a state of constant personal crisis about the economic solutions inherent in the green economy, then they get excited. Nowhere is an eco-populist, opportunity-based message more important than in engaging people of color.

As pointed out above, we should advance more popular slogans that present green solutions to real-life, kitchen-table problems. I have discovered some with real appeal and resonance: "Green the Ghetto" and "Green-Collar Jobs for All" (or "Green Homes for All," or "Solar for All," or "Organic Food for All"). I think we should explore a clean-energy call framed as the desire for "Asthma-free Cities."

One urban eco-populist slogan stands out above the rest, with a power all its own. It speaks to the full range of urban concerns, addressing simultaneously issues of economic justice, criminal justice, and environmental justice. That slogan is "Green Jobs, Not Jails." My hope is that it will someday be adopted and embraced by the entire green movement as the central goal guiding our efforts.

The point is that these eco-populist slogans—and programs to back them up—will be key to engaging people of color and other disadvantaged communities in the struggle for a green-collar economy. To those who have plenty of personal opportunity, speak first about the environmental crisis. But to those who have plenty of personal crises, speak first about the environmental opportunities—and how solutions for the Earth's woes can be solutions for their problems too.

THE NOAH PRINCIPLES

So the direction is set. We seek a social-uplift strategy that creates green jobs, not jails; a politics anchored in a Green Growth Alliance for this century; and a moral framework based on reverence for each other and the planet. It can be done. However, there are many habits of mind and unconscious assumptions that stand in the way, even for those of us who have been lifelong change makers. We need some new distinctions and ways of seeing our work, or we are likely to reproduce some of the same negative patterns in this movement that we have seen in others.

For instance, when I was a young activist, the role model I subconsciously had for making a difference was David confronting Goliath. And that image or archetype served me well—for a while. After all, the David and Goliath story is a beautiful tale, one that foregrounds courage and allows for the possibility of miraculous outcomes, of defeating the bully against all odds. I have come to believe now that there is also a shadow side to the myth. It requires that the protagonist always be small and marginalized, and it requires a politics of confrontation and opposition. Such a politics may serve us poorly as we confront the dangers that will demand cooperation on a massive scale.

So I raise the possibility that we need a new guiding narrative, a new myth, for the new challenges that face us. Our leaders need a different yet familiar story that defines the kind of leadership we need.

With violent storms, rising seas, and financial chaos darkening the horizon, perhaps the best models for the new century will prove to be Noah and his wife. Theirs is a story of leaders who must make plans for a difficult future while trying to save as many people and fellow species as possible. It's also a story about honoring and managing diversity, about making space for everyone from the tiniest termites to the lions and the elephants. Instead of preparing to protest against a giant, as David did, perhaps it is better to prepare to lead a community through a crisis and into the future beyond that crisis, as Noah did.

Of course, there are good people on the other side of important disputes who will not be won over; they will have to be run over. There will be times when we have to fight—in the old "us versus them" mode. But when we do, we want the biggest possible "us"— and the smallest possible "them." I mean, when all life everywhere is threatened, we might even need Goliath to help us build the ark.

A politics in keeping with Noah's principles would focus on creating something new rather than confronting something old. It would be more about "proposition" and less about "opposition." As guideposts to creating that kind of politics, we could advance the following five points. Call them the Noah principles:

1. Fewer "issues," more solutions

2. Fewer "demands," more goals

3. Fewer "targets," more partners

4. Less "accusation," more confession

5. Less "cheap patriotism," more deep patriotism

Fewer "Issues," More Solutions

Organizations working for change usually place themselves into one of two categories: single-issue groups (e.g., fighting against homelessness) or multi-issue groups (e.g., fighting against police abuse, prison expansion, and youth violence). I propose a different distinction: issue-based groups and solution-oriented groups. After all, the word "issue" is just another word for "problem." If you have an "issue-based" group or coalition, you essentially have a "problem-based" organization. And defining any cause based on a negative can lead to a great deal of negativity.

If you doubt me, try this experiment. Approach almost any hard-working activist committed to a cause and say: "Tell me what issue you are working on." The activist will talk to you for an hour, pouring out all the horror stories, pet anecdotes, and shocking statistics that animate and inform her or his work. The minute the activist runs out of breath and you see a chance to get a word in—seize it. Say, "Okay, so tell me what solution you are working for." Most of the time the person will fall silent and then perhaps start yammering and stammering. But I can almost guarantee you that the problem statement will be much sharper and generate much more passion than the solution statement.

Eco-heroine Julia Butterfly Hill has an explanation for this phenomenon. She says of people who have committed themselves to important causes: "Many of us have gotten so good at defining what we are against, that what we are against has started to define us."

And it is true. Many individuals and organizations define themselves solely by what they are against. They are anti-racism, anti-sexism, anti-homophobia, anti-globalization, anti-imperialism, anti-capitalism, anti-corporation, or anti-war. Many of us on the left define ourselves in wholly negative terms, and then we wonder why people run the other way when they see us coming.

As we build new organizations and networks, it is not enough to know what we are against. Saving the Earth and its peoples requires that we also know—and know with specificity—what it is that we are for.

Fewer "Demands," More Goals

Some people who want to use their talents to make change, rather than money, decide to become community organizers. They sign up with a community-based organization or an environmental campaigning group. And then they go to workshops or even training camps to learn how to do their jobs right.

Invariably, they are taught some version of a three-step process. The organizer is supposed to help aggrieved people come up with a list of "demands," help pick a "target" who can meet those demands, and go with others to "pressure the target" into meeting their demands. This approach constitutes the basic, underlying "operating system" for activism in the United States. It is the literal codification of the David and Goliath approach to social change: the little "we" versus the big "them."

The truth is sometimes that formula works. And sometimes there is no obvious alternative. And yet the problem with always formulating our desires as "demands" is that the word itself assumes (and may reinforce) an adversarial relationship. As the *Titanic* sinks, one has to wonder whether multiplying adversaries is a very good idea— especially for those trapped at the bottom of the ship, who may need more allies than enemies.

One simple option is to reimagine and reformulate our desires as "goals." Goals can be shared—even by people who disagree on many points. Demands can never be shared. One party makes them; the other party must either deny them or capitulate. A victory under those circumstances can feel quite hollow. Sometimes we can win the short-term battle, but lose the long-term aims.

I have often wondered how far I would get if I marched into the offices of some social change groups and made "demands." I don't think they would be very receptive. Sometimes a mere change of language can change the mind-set—on both sides—and possibly yield a much greater outcome. Given the threats we face, it should be worth a try.

Fewer "Targets," More Partners

The other problem with the standard "organizer's" formulation is the constant seeking out of "targets" to pressure. Again, sometimes there is a person in a position of authority who is so obstinate, so biased, and so recalcitrant that one has no choice but to declare him or her an opponent. Also, there are institutions that have acted in bad faith for so long that trust is almost impossible to regenerate. Under those circumstances, the advocates of a righteous cause do need a battle-hardened cadre of well-trained organizers who know how to twist pinkies and otherwise force an adversary into submission. The protectors of the status quo use power tactics all the time. The champions of a better tomorrow should not unilaterally disarm.

Yet this approach can be overdone, overplayed, and overused. Too often activists just assume that any change worth making will always require a big battle with someone. They start preparing for Armageddon every time any issue comes up—even before they have taken the first steps to resolve it using less confrontational means.

Some organizations are like countries run by generals who have an army but no diplomatic corps. Therefore, they spend all their time drilling their troops and scanning the horizon, hoping for an opportunity to declare war on someone or something. Again, sometimes this is justifiable.

However, it comes down to a question of balance. When all you have is a hammer, everything looks like a nail. If all you have is "direct-action organizing," everyone with power looks like a target. It does take real skill, talent, and training to identify targets, challenge them, and get them to do what you want. The time has come for social change and environmental organizations to add another set of skills: the ability to turn would-be "targets" into real, long-term partners for change. And that can be a tougher challenge.

For one thing, it requires that organizers move beyond assumptions, stereotypes, and past hurts. It requires that organizers (and those they organize) invest time in relationship building and trust building across lines of race, class, and authority—trying to surface points of shared interest and concern. Not every grassroots group has the time, capacity, or organizational strength to function in this way. Sometimes it is easier for marginalized activists to just call a press conference and start painting protest signs. I understand that. Yet over the long term, the accumulated results often are not worth all the expended effort.

Here's the truth. If you rush into a situation looking for enemies, you will always find plenty. At the same time, if you go into a situation trying to find friends and allies, you will almost always find at least one. Sometimes they are in surprisingly powerful places—like behind the receptionist's desk in your opponent's front office.

In this age, our main job is to seek out friends wherever we can, not just to defeat enemies. What if, rather than mainly looking for opponents to punish, those of us who are committed to social change spent our time seeking out potential allies to encourage, befriend, and reward. After all, for every scofflaw polluter, there may be dozens of local businesses out there trying to do the right thing ecologically—but getting little support or recognition. What if environmentalists did more to partner with them, celebrate them, and help them? For every racist employer or bigoted beat cop, there are tens of thousands of white people who absolutely abhor racism. And yet civil rights activists like myself rarely ask them to do anything—except to feel guilty. Why not focus on finding better ways to access their time and talent for the good of all?

Our cause needs fewer enemies and more friends. To get through the coming crises, we are going to need each other. Let's start laying the groundwork now, so that—later on—we will be more likely to turn *to* each other, not *on* each other.

Less "Accusation," More Confession

None of this is to say that we won't have to confront and defeat real, implacable, and unyielding enemies on our journey. We will. If we could achieve eco-equity only by defeating external enemies, we would be walking an easier path than the one we're on.

That's because some of the enemies we need to defeat are inside us. We ourselves are a part of the problem. Every day, almost all of us are working and consuming in the pollution-based economy. We are participating in an economy that lacks equity, and yet we each have an understandable aversion to giving up our own money or status. We are trying to change the status quo. But we all have stake in it too. We all rely upon it to live and survive. And so, every day, we end up feeding the very monster we are fighting.

This is a humbling fact. If we are honest, even those of us who desperately want change must admit that we are not just battling the polluter without; we are also battling the polluter within. We can say this not just in the obvious "material" sense. This is not just about how many times we fail to recycle, bicycle, or bring our cloth bag to the grocery store. We all have inner demons that pollute our minds and hearts—that cloud our thoughts and distort our actions.

As we try to work with others, our egos often get in the way. Tempers flare. Suddenly we find ourselves not just battling the warmonger without, but that white-hot warmonger within. Later on, resentment creeps in over some perceived slight and indignity. Eventually, we find ourselves battling not just the punitive, unforgiving jailers on the outside. We end up battling the punitive jailer within, the hurt and angry part of ourselves who can't forgive our coworkers and allies for shortcomings and disappointments. These are the hidden struggles that define our days. Cumulatively, these inner tumults determine and limit the impact of our work itself, but nobody talks about them much.

Instead, we engage in the old politics, naming, blaming, and shaming somebody else while concealing our own faults, flaws, and hypocrisies. However, the cause of pursuing eco-equity does not easily lend itself to that approach. The change we are seeking is too monumental, and our own capacities are too modest.

We would be better off confessing our own weaknesses, our fears, our needs. Doing so will let others see the gaps more quickly, find their rightful places around the growing circle—and come to the campfire with fewer pretenses themselves. If we confess our own struggles to realign our own lives and change our own behavior, we may seem less alien to those we are trying to convince.

Also, the change we seek is so complex that no one person can understand everything that must be done. In that regard, we are all equally ignorant about how to get where we are going. This weakness actually is our strength. If we confess our own uncertainty, we are much more likely to listen attentively to others—and pull others into speaking more honestly and fully. As we move forward, our motto should be: accuse less, confess more.

Less "Cheap Patriotism," More Deep Patriotism

We have gone through a period during which people waving American flags have done great damage to the country, to the people of Iraq, to America's prestige in the world, to the national treasury, to the U.S. Constitution, and to the international rule of law. While force-feeding the country a brand of cheap and mindless patriotism, the "leaders" waving the biggest flags have steered the nation into a ditch. People of conscience should embrace Old Glory—and use the flag to help guide the public back in the direction of sanity.

One begins to fear that this accident was not very accidental. After all, GOP anti-tax operative Grover Norquist had declared openly: "I don't want to abolish government. I simply want to reduce it to the size where I can drag it into the bathroom and drown it in the bathtub."[8] That is not a patriotic statement.

We have an obligation to tell the ultraconservatives who are so rabidly antigovernment: "If you don't love this government, then let it go and hand it over to people who do." Those who would hijack the government and crash it with deficits pose a bigger threat than the terrorists.

And while we are at it, we could make do with a lot less knee-jerk antipatriotism from the left. I know it is hard to make peace with the country's original sins of stolen land and stolen labor. It is hard to forgive its repeated entanglement in unjust wars, up to the present moment. However, the far left's strategy of trying to fix the country by putting it down all the time has been an utter failure.

To paraphrase scholar Cornel West, you can't save a country you don't serve, and you can't lead a country you don't love. And there is much to love in this country. After all, we are talking about the nation that gave the world basketball, iPods, and Beyonce Knowles. (If those three won't get you up stomping and cheering for the red, white, and blue, I don't know what will.)

The United States has the power to be a huge obstacle to planetary survival—or giant springboard to planetary salvation. A better America is the best gift that we can offer the world. Yet caring Americans will never give the world that gift if they are holding their noses and handling the flag like a used tissue.

If we do our work right, the United States will lead the world, again, someday. This next time—not in war. Not in per capita greenhouse-gas emissions. Not in incarceration rates. The United States will lead the world in green economic development, in world-saving technologies, in human rights. We will lead by showing a multiracial, multifaith, rainbow-colored planet how our multiracial, multifaith, rainbow-colored country pulled together to solve tough problems. The United States will go from being the world leader in ecological pollution to the world leader in ecological solutions.

Bruce Springsteen put it best in 2004 when he said: "America is not always right. That's a fairy tale for children. . . . But one thing America should always be is true. And it's in seeking her truth, both the good and the bad, that we find a deeper patriotism, that we find a more authentic experience as citizens, that we find the power that is embedded only in truth to change our world for the better."[9]

It's time for the deeply patriotic to take back the flag from the cheaply patriotic, because, despite the pain of old crimes and recent disappointments, some of us still believe in America. Some of us still believe in "a more perfect union"—and in making it more perfect every day. Some of us still believe in "America the beautiful"— and in defending its beauty from the clear-cutters and despoilers. Some of us still believe in "one nation, indivisible"—and in opposing those who profit by keeping us needlessly divided. Some of us still believe in "liberty and justice for all," and we won't stop until that classroom pledge is honored from shore to shore.

Some of us still believe in America—and in all of those things we learned about it as children. Of course, we know now that America is not the place we live, but a destination to which we all are headed. So we keep faith on the journey. No, some of us haven't given up on Dr. King's dream. There are those of us who yet believe we are going to win.

And when we do, we'll be doing more than just "taking America back." We will be taking America—forward.

FIVE

The Future Is Now

W E HAVE EXPLORED the principles and the politics that
could revive the economy on a more inclusive and ecologically responsible basis. Luckily for us, leaders in far-flung and
unlikely places are already moving ahead and creating this future.
They are not waiting for federal action to give them the green light
to start creating the new economy.

We will examine courageous pioneers who are already helping ordinary people blaze green pathways to prosperity. I hope that someday the vast majority of U.S. workers will have jobs in the kinds of
innovative enterprises and programs that we explore below. After
all, every day, about 145 million people go to work in the United
States.[1] Imagine if those jobs—plus new ones created for people who
are currently unemployed—were largely working in fields and pro-
fessions that uplift human dignity and honor the Earth.

Some are already doing it today. Here are their stories, organized
by five major subsystems of sustainability: energy, food, waste,
water, and transportation.

ENERGY

The transition from our reliance on fossil fuels to clean and renewable energy is the linchpin of the green economy. If done correctly, it will bring our carbon emissions down to a manageable level. It will free us from foreign oil and its national security risks. It will halt the skyrocketing rates of pollution-based illnesses, and it will revitalize our economy and create millions of green-collar jobs.[2] Here are projects that are simultaneously producing clean energy and creating career paths for the unemployed.

Energy Efficiency

The cleanest energy is the energy that we never have to use—because we were wise enough to conserve it. We waste a lot—in heating, cooling, and lighting our drafty, poorly designed homes and offices. Not to mention driving our cars, with their unimpressive number of miles per gallon. The energy we don't use is cheap, silent, clean. It's measured in "negawatts," instead of megawatts. Improvements in the energy efficiency of buildings (such as weather stripping, replacing fixtures, and insulating hot-water heaters) can simultaneously save property owners money, reduce demand for fossil-fuel-generated electricity, and provide both skilled and unskilled jobs for local workers.[3]

In Los Angeles, the community-based group Strategic Concepts in Organizing and Policy Education (SCOPE) convened the local Apollo Alliance. Campaign Coordinator at SCOPE, Elsa Barboza says the local Alliance's first step was "to collect signatures from black, Latino, Asian, and Anglo working-class families throughout Los Angeles' inner-city neighborhoods for a petition to create a sustainable, equitable, and clean energy economy that will bring quality jobs to their communities, create a healthier and safer environment, and promote community-based land use planning and economic development."[4]

One of the people out knocking on doors with the petition was Oreatha Ensley, a lifelong civil rights activist, a mother and grandmother, a former teacher, and an LA resident for nearly forty years. She says: "I expected some folks to tell me that jobs are number one and cleaning our environment is just a nice wish. Instead, they told me that it's about time we reinvest in our community, because we are slipping away further into poverty and getting sicker because of it."[5]

It wasn't just the poor communities who were in favor of the Alliance's objectives. "Mayor Villaraigosa, a liberal-leaning City Council, and forward-thinking Commissioners have articulated a bold vision to make Los Angeles a national leader in the transition to a sustainable, equitable, clean energy economy," Elsa notes. Indeed, the City of LA has already committed to one of the local Apollo Alliance's proposals: implementation of a pilot program to retrofit one hundred city-owned buildings with energy- and water-conservation technologies.[6]

Nationwide, buildings are responsible for 36 percent of our energy use, 30 percent of our greenhouse-gas emissions, and 30 percent of our waste production. Once complete, retrofits in LA are slated to save the city up to $10 million per year in utility costs.[7]

The City of LA owns and operates more than eleven hundred buildings, many in deteriorating condition, that cover over a million square feet. The work includes audits; energy-efficiency improvements (e.g., sealing around or replacing doors and windows); lighting upgrades (replacing bulbs, installing sensors, and maximizing daylight); water-conservation improvements (fixing leaks, replacing urinals and toilets); healing- and cooling-system updates; and cool- or green-roofing installation Audits—the first step in the process—are under way in LA as I write this.[8]

In the longer term, workers from the retrofit jobs can be transitioned into maintenance and construction jobs in both public and private sectors. According to the California Employment Development Department, employment opportunities in construction in the LA area are projected to increase by 30 percent between 2002 and 2012. The industry's career ladders allow workers earning entry-level wages of $9 to $18 per hour to become advanced skilled workers such as plumbers, sheet-metal workers, and electricians, earning $15 to as much as $50 per hour.[9]

Meanwhile, in the old industrial city of Milwaukee, Wisconsin, an organization called the Center on Wisconsin Strategy (COWS) is exploring how workers in the Rust Belt can move to the center of the clean-energy economy. COWS has cooked up a brilliant scheme to retrofit all of Milwaukee's buildings—and to create a slew of green-collar jobs in the process. To make this work, a new building-efficiency service—Milwaukee Energy Efficiency (Me2)—is being created to offer all of that city's residents the opportunity to buy and install cost-effective energy-efficiency measures in their homes and businesses.

Here's the beautiful part: there will be no up-front payment, no new debt obligation. Customers will have full assurance that their utility costs will be lower, and they will make monthly payments only for as long as they remain at the location and the measures continue to work. Outside of *The Godfather,* that's about as close as you can get in this life to "an offer they can't refuse."

This is how it will work. Owners or renters sign up to have their place retrofitted to save energy costs; a qualified person shows up and does the work; and then the renters or owners pay off the cost of the retrofit a little bit at a time, over the course of years, as a part of their (now radically reduced) electricity or property-services bill. That's it. Everybody wins—including the Earth.

In the policy section that follows, we will go into more detail about how this program works. For now, suffice it to say that the

Me2 program will be great for those community residents seeking jobs. COWS and the University of Florida estimate that every $1 million spent on the effort will generate about ten job-years in installation and construction activities and another three job-years in upstream manufacturing for needed parts. For a roughly $500 million project, that's a lot of job-years: about sixty-five hundred. Residents can start with less-skilled work—like blowing insulation or wrapping pipes—and move up to the more advanced work in plumbing, wiring, and installing new heating and ventilation systems.[10]

Retrofitting work will provide good jobs—family-supporting gigs, with solid opportunities for advancement—that cannot be outsourced. They will feed into the exploding green building industry, and that will simultaneously reduce our emissions and our reliance on foreign oil. Similar energy-efficiency programs should be implemented across the country.

Wind Energy

Winona LaDuke cites a prophecy among her people: "In the future, we've got two paths ahead. One path is well-worn . . . but it's scorched. The other path, they say, is not well-worn, but it's green. It's our choice upon which path to embark."[11] LaDuke is a member of the Anishinaabe tribe, a Native activist, and twice the Green Party candidate for vice president on the ticket with Ralph Nader. She has chosen her path and has spent much of the last decade rallying other Native Americans to embrace the green economy. It's not such a tough argument to make to her people, when the economic implications for Indians are considered.

"Native Americans are the poorest people in the country," notes LaDuke. "Four out of 10 of the poorest counties in the nation are on Indian reservations. . . . Over half of the American Indians on my reservation live in poverty. . . . Unemployment on the reservation is at 49 percent according to recent BIA statistics. And nearly one-third of all Indians on the reservation have not attained a high-school diploma."[12]

At the same time, Native America contains much of the continent's old-school energy resources. One-third of all uranium and two-thirds of all low-sulfur coal come from Native lands. The largest coal strip mine in the world is on a Native reservation. Massive hydroelectric projects in the subarctic are also on Native lands.[13]

One might assume this is good news for Native America. But Indians suffer from the disastrous and toxic side effects of extraction of these resources. Mining operations cause the displacement of communities, destruction of natural habitat, disruption of sacred sites, and water pollution with deadly toxins. Industrial toxic and radioactive wastes accumulate in fish, crops, animals, and soil. Oil drilling and related activities fragment, deforest, and pollute the landscape and fragile ecosystems. Clear-cutting and other intensive logging methods destroy the habitat of animals and fish, cause soil erosion and thermal pollution, and pollute water with both sediment and herbicides. Large hydroelectric projects flood lands needed for crops, disrupt and destroy subsistence-based cultural practices, and forcibly displace entire communities.[14]

Trying to get rich by further devastating the land has little appeal for those committed to traditional ways. Fortunately, it turns out that Native lands are also extraordinarily rich in clean energy, wind in particular. The wind potential of twelve reservations in North and South Dakota alone could meet 41 percent of the U.S. energy demand, and more than half the country's electricity demand could be met by the wind available on all the reservations.

The first turbine on Native lands was installed in early 2003 on the Rosebud Sioux Tribe Reservation in South Dakota. It produces enough clean electricity each year to power over two hundred homes. Another was erected on the Mandan, Hidatsa, and Arikara

Nations' reservation in North Dakota in 2005; six more were set up in 2006 in the Native villages of Toksook Bay and Kasigluk, Alaska (three per village); and many more are under construction. Rosebud alone aims to produce 50 megawatts by 2010.[15]

That's environmental justice—and poetic justice—for you. And there's more poetry in a former steel town in Pennsylvania.

When the U.S. Steel plant on a three-thousand-acre site along the Delaware River in Bucks County, Pennsylvania, shut down the last of its operations in 1991, it brought nearly forty years of manufacturing-based stability to an end. At the height of its operations, in the mid-1970s, the plant had employed more than eight thousand people. "This is a sad day for U.S. Steel, our employees and the communities surrounding," Thomas J. Usher, chairman and chief executive officer of U.S. Steel's parent, USX Corporation, said at the time.[16] Little could he imagine that, just a little over a decade later, the site and some former steelworkers would be in the business of manufacturing components for clean-energy projects.

In 2005, the Spanish wind-energy company Gamesa bought twenty-four acres of that plant. Gamesa is among the largest wind-energy companies in the world. Currently it is the only vertically integrated wind-energy company in the world, meaning it manufactures the parts for wind-energy units and also develops wind farms itself.[17]

Pennsylvania governor Ed Rendell and the head of Pennsylvania's Department of Environmental Protection, Katie McGinty, wooed the Spanish company. They were eager to attract the manufacturing jobs and meet the state's Alternative Energy Portfolio Standard, which dictates that 18 percent of the state's power come from renewable sources by 2020. The state extended Gamesa $10 million in grants, loans, and tax credits from the state's Department of Community and Economic Development and the Bucks County Economic Development Corporation. Because the site is a brownfield (a "Keystone Opportunity Improvement Zone," in Pennsylvania's terms), Gamesa gets breaks on state and local taxes through 2019.[18]

One of Gamesa's thirteen hundred workers is former U.S. Steel worker Jim Bauer. He had been a crane operator for twenty-five years when U.S. Steel laid him off. Now he's back. At Gamesa, he heads up a team that assembles parts for the turbines, and earns $17 per hour, just slightly less than his U.S. Steel wage. The wind workers are even part of the same union they belonged to in the old days, after United Steelworkers of America persuaded Gamesa not to fight an organizing drive. Jim says he's proud of his new work making clean energy—proud to be keeping America safe and free from foreign oil and keeping the planet intact for his children.[19]

Solar Energy

Richmond, California, has been in need of some healing. Its motto is "The City of Pride and Purpose," but in recent decades it's been better known as the city of *violence* and *pollution*. In 2004, it was ranked the eighth most dangerous city in the country, and the second most dangerous in California after Compton and ahead of Oakland. The following year, city-council members declared a state of emergency due to the crime rate.[20]

And the pollution? Chevron USA has a major oil refinery in the city, with a storage capacity of fifteen million barrels. The refinery often releases toxic gases and has had many disastrous chemical spills and leaks, often of chlorine and sulfur trioxide. Driving back from pristine Marin County just on the other side of the Richmond–San Rafael bridge, I've had to roll my windows up more than once to blunt Chevron's stench.

Under the leadership of a Green Party mayor, in 2007 Richmond joined together with neighboring Berkeley, Emeryville, and Oakland to form the East Bay Green Corridor. Mayor Gayle McLaughlin said at the time: "Given the crises the world faces from resource depletion, poverty, and species extinction, a green economy is also the only way to reinvigorate our economy while at the same time addressing environmental destruction and social inequity. Cities generate 75 percent of the carbon emissions. Cities like ours are where the problem must be solved."[21]

Lighting the way in Richmond is a solar-energy initiative organized by a group called Solar Richmond, a coordinating organization that is partnering with solar-panel vendors, city agency RichmondBUILD, and nonprofits Solar Living Institute and Grid Alternatives. Together they aim to create a hundred new family-supporting solar jobs and achieve 5 megawatts of solar energy in Richmond by 2010. The founder and director of Solar Richmond, Michele McGeoy, is on a mission to restore the pride and purpose of its slogan to the city. "Solar is one antidote to pollution, and jobs are one antidote to violence," she says.[22]

In 2006, McGeoy contacted Sal Vaca, Director of Richmond-WORKS, the City's employment and training program. She persuaded him to send three students in the City's vocational program in construction trades, RichmondBUILD, to her home and allow them to spend three days learning—on the job—to install solar panels as a pilot project. The following week, her newly installed panels were slated to be part of a solar tour as a shining example of what Richmond's young people could do in the promising field.

In a great example of innovative civic partnerships, installations for low-income homeowners are made possible through the City's redevelopment agency, which grants loans for the equipment to these homeowners, at cost. The labor by RichmondBUILD trainees for the installations is free. If the homeowners are senior citizens or disabled, it's a deferred loan. The loan for the equipment is perhaps $6,000, but it increases the value of their home by about $12,000. Plus it lowers their utilities bill for the rest of their lives.

Angela Greene is one of the graduates of the training program. Now forty-seven, Greene spent twenty years in the printing industry, working her way up from being a messenger to managing her own store. A single mother of two girls, she was nevertheless able to buy her family a house in Richmond. And then the parent printing company with which she was affiliated shut down, and Greene lost her job. She struggled to make ends meet while on unemployment, had to give up her car, and fought to hold on to her house.

Then she heard that the City of Richmond was offering a construction training program. Figuring she would at least learn skills to save her money around her own home, she enrolled. That's how she was introduced to McGeoy and Solar Richmond.

"I'd never really thought about carbon dioxide emissions before that," Greene says. "I'd been doing recycling and gardening, but the solar training started really making me think about what we were doing to the Earth. I want my children to be on this Earth to see their children." She says she saw the dot-com boom come and go without benefiting from it—this time she was determined to be a part of the boom.[23]

There are a lot of opportunities for graduates. Solar Richmond's immediate goal is to get graduates placed into entry-level installer positions. From there, they can become a project lead, and from there, a project manager—assuming more supervisory roles. McGeoy says, "Right now, the industry's biggest shortage is project leads. What's exciting is that the field is so new that you only need two to three years of experience to be considered an expert. Compare that to the trajectory of a traditional electrician, where you have to have fifteen years under your belt to be considered experienced. That means that three to five years from now, our graduates are going to be doing a lot of the management. The demand is there." In fact, in California, the demand for expertise in the field of solar power distinctly outstrips the supply.[24]

Another graduate of the RichmondBUILD/Solar Richmond training program is Rodney Lee, who found his way to the program after losing his job at the local telephone company due to an extended illness. He had never seen a solar panel before starting the program. Now Lee's an operations assistant at a Silicon Valley–based company called SolarCity.[25]

Founded in 2006 by two brothers, Lyndon and Peter Rive, Solar-City became the fastest-growing solar-power company in California in less than two year's time, with over 250 employees at time of publication and $29 million in sales in 2007. The company expanded to fellow solar-friendly states Oregon and Arizona in 2008 and expects to be on the East Coast soon.[26]

The Rive brothers quickly realized that the major barrier to solar adoption was the prohibitive cost of equipment and installation—anywhere from $20,000 to $100,000 or more. They knew that a successful solar business model would need to be built on making solar more affordable. They felt that if they could convince an entire neighborhood or community to go solar, they could lower the cost for every resident. They even created a financing program called SolarLease, backed by Morgan Stanley, to help more people get panels.[27] As far more residents go solar—and entire communities in some cases—SolarCity has been able to create more installation jobs.[28]

FOOD

At first glance, the food system—how we fuel our bodies—may seem less connected to climate change than how we fuel our cars or how we heat, light, and cool our homes. Yet consider just one random and bizarre fact that drives home the interconnectedness of our systems: as a result of the ten pounds that the average U.S. citizen gained in the 1990s, the airline industry has burned 350 million additional gallons of fuel per year.[29] Although that's among the more obscure correlations you can find, our food system is a major consumer of precious resources and has, between its production processes and global distribution, a significant carbon footprint.

Trade and food-security expert Anuradha Mittal comments on the ridiculous distances that food travels in fossil-fueled transport: "Today 20 percent of California table grapes go to China, while China is the world's largest producer of table grapes. Half of all California's processed tomatoes go to Canada, and the U.S. imports $36 million worth of Canadian processed tomatoes yearly. . . . We are exporting what we are also importing because it is profitable for the companies doing it, not because it is good for the nation or the environment."[30]

The real and often hidden costs of the food system include water pollution from chemical runoff and factory farm waste, poisonings caused by pesticides, greenhouse-gas emissions from livestock and cropland operations, the loss of topsoil and soil deterioration, the operation of multiple federal agencies tasked with regulating the food industry and mitigating damages, federal farm subsidies, the use of oil to develop petrochemical fertilizers and pesticides, the use of petroleum in food processing and transportation, and the use of preservatives and sometimes packaging needed for foods traveling long distances. One estimate places the total of these true costs at $40 billion per year.[31]

Meanwhile, people in our own country are actually going hungry. According to the U.S. Department of Agriculture (USDA), in 2006, 35.5 million Americans lived in "food insecure households," comprising 22.8 million adults and 12.6 million children.[32] Many low-income communities, particularly in urban settings, don't have access to a single real supermarket. They are forced to patronize liquor stores selling Cheetos and Snickers—if they're lucky maybe a potato or a banana—all at 30–70 percent higher prices than regular stores.[33] And of course there are the ubiquitous fast-food chains. Medical costs to treat diet-related ailments like diabetes and heart disease run more than $75 billion a year in the United States.[34]

A final failure of the system is the severe negative impacts on the people who work in the food industry, particularly farm laborers. Nearly seven out of ten U.S. farmworkers are foreign-born, 94 percent of them from Mexico.[35] Although there are devices available that lessen the physical impacts, many workers still engage in backbreaking "stoop labor," as César Chávez termed it, when farm owners won't pay for appropriate ergonomic tools. Our current system also deeply dishonors the work of traditional farmers, who are not only going out of business at record rates—more than seventeen thousand each year, or one farmer every half hour—they are also killing themselves at record rates; suicide is now the number-one cause of death among American farmers.[36]

It is clear that the system has to be reformed. There needs to be a shift to local food that is organic, free of pesticides, fertilizers, and preservatives; that is produced in ways that do not harm consumers, food workers, animals, or our soil and water; and that provides dignity to those who produce as well as consume it.

Happily, a movement for community food security is growing. This movement holds that a local and sustainable food system can not only guarantee nutrition and health; it can maximize community self-reliance and social justice. It can create jobs. Urban and peri-urban (on the edges of cities) agricultural projects are opening new labor markets. And these can be green-collar jobs.

Two of the fiercest catalysts for change in the arena of food-system reform and organic urban agriculture initiatives are LaDonna Redmond, in Chicago, and Brahm Ahmadi, who heads up the People's Grocery in Oakland. Oakland might seem logical. It is so close to the cradle of fresh and seasonal food (as represented by Chef Alice Waters in Berkeley). But Chicago?

LaDonna Redmond's involvement in urban agriculture began in 1999 after her infant son Wade was diagnosed with severe food allergies to a range of items from peanuts and shellfish to eggs, cheese, and milk as well as a host of additives. As a concerned mother, she started researching allergies and, to her horror, learned about the chemicals involved in standard food production: GMOs (genetically modified organisms), preservatives, additives, growth hormones, pesticides, and fertilizers (including toxic sewage sludge). She decided that Wade—and the rest of her family—needed a whole-foods diet with as little processed, packaged food as possible: "I needed to gain access to food unpolluted by genetic engineering and free from pesticides. I needed organic food."

That was easier said than accomplished. Her search for organic food took her out of her Westside neighborhood and through the city of Chicago; her search for *affordable* organic food was entirely in vain. "You could buy two-hundred-dollar sneakers, semiautomatic weapons, and heroin, but you couldn't get an organic tomato," she said. The idea that her community did not desire high-quality organic food was one of the many myths she wound up shattering.

Indeed, once LaDonna and her husband, Tracey, turned their backyard into an urban "micro-farm," as they called it, neighbors started coming by to help and share the harvests. Lettuce, tomatoes, peas, squash, greens, cabbage, onions, and herbs were all part of the early abundance, with more than enough to share. One thing led to another, and today the Redmonds' organization, the Institute for Community Resource Development, secures empty lots from the city, oversees a whole network of lots-turned-gardens, manages a farmers market, provides technical support and nutritional education, and is planning for the opening of a retail store.[37]

City support has been crucial. Income from food sales can pay for maintenance and staff salaries, but start-up costs can be prohibitive, especially since land values in urban centers such as Chicago are relatively costly. Luckily Chicago's Mayor Daly is committed to leaving no stone unturned in the greening of his city.

"LaDonna's projects go beyond mere gardening because the intent is to look at the comprehensive approach to the issue of developing local economies—hiring locally, selling locally," notes fellow Chicagoan Orrin Williams, founder and director of the Center for Urban Transformation. In the eighty thousand vacant lots totaling several square miles, plus the thousands of flat rooftops and backyards, Williams guesses that as much as 40–50 percent of the city's food could be grown. Another estimate puts the number of full-time jobs that would result from cultivating that land at forty-two thousand.[38]

Although it's been the construction trades that have traditionally been most receptive to hiring formerly incarcerated people, agriculture presents opportunities for them as well. Some prisons even smooth the path by offering horticultural training programs. (In the Bay Area, Catherine Sneed's "garden project" at the San Mateo jail still sets the gold standard for this kind of work.)

Williams also works as the employment training coordinator for another organization in Chicago called Growing Home, which fosters life and job skills in a transitional employment program for previously incarcerated, homeless, and low-income Chicagoans. Many are recovering from addiction, suffering from mental illness, or have not held a steady job in years. But more than 65 percent of the folks who go through Growing Home's transitional job-readiness program find full-time work in the retail, landscaping, and food-service industries or placement in further training or educational programs.[39]

Working in the soil is life-affirming. One Growing Home graduate who found the program after spending three years in a correctional facility for using and selling heroin and cocaine, says she "loves seeing little sprouts push up through the ground."

George Washington Carver wisely said: "I believe that the great Creator has put ores and oil on this earth to give us a breathing spell.

As we exhaust them, we must be prepared to fall back on our farms, which is God's true storehouse and can never be exhausted. We can learn to synthesize material for every human need from things that grow."[40] When it is grown and processed in ways that do not harm the Earth or its inhabitants, there is great dignity to the creation of food.

Back in Oakland, Brahm Ahmadi shares this core belief. He's the son of a Midwestern mother descended from generations of farmers in the Iowa-Missouri area and an Iranian father from merchant families in Tehran and Tabriz. Ahmadi understands the deep significance of food and simultaneously believes in the potential for entrepreneurship to create innovative solutions.

Ahmadi started out in the environmental justice community, defending the poorest communities from toxins. Yet that work took its toll. Ahmadi realized that he and his colleagues were good at articulating the problems and being angry about them, but they fell short on inspiring hope and possibility.

So now, as the executive director of the People's Grocery, he's all about solutions, particularly as applied to the community of West Oakland. It's a place facing multiple challenges, where residents' economic as well as mental, emotional, and physical health issues compound one another. The lack of amenities and services (which fled to the suburbs), the police brutality, the environmental toxins—in West Oakland all these have converged to create a crisis situation. Chronic diseases are at epidemic levels. There are severe mental-health challenges. Overall, there are a lot of struggling families and individuals.[41]

"Food is our medium for achieving broader outcomes in community development and public health and addressing disparities in opportunities and quality of life," says Ahmadi. "We chose food as our tool because it's intimate and universal, regardless of differences in culture or personal preferences. On a fundamental level, we all have to eat every day, and we have that in common."

"We've taken risks, in the entrepreneurial tradition, which isn't as common in the nonprofit world," says Ahmadi. The risks are paying off. People's Grocery has expanded from its three urban gardens in Oakland and signature Mobile Market to a two-acre farm in nearby Sunol, growing nearly eighteen thousand pounds of produce. These enterprises not only provide healthy foods and good jobs; they also educate community members.

Ahmadi says:

We get started with a conversation about individual food consumption, the meaning of what you eat, and the history behind why certain food is or isn't available to you. From there we connect the dots to the structural and systemic issues of the food system: considering the global environmental footprint of food production, how far food travels, and equity issues related to farmworkers and the struggles of small farmers . . . connecting those to the struggles of low-income urban consumers.

Despite its public image, there is significant spending power in the community. People's Grocery has assessed West Oakland as an approximately $50 million food market, of which about 70 percent is not captured locally. That's a lot of money being sent outside the neighborhood and thus not contributing to local jobs and wealth. Ahmadi dreams of:

completely localized food systems that are regionally based, with the majority of the food that consumers consume coming from within a few hundred miles of where they live, so that consumers have direct knowledge of the farms and farmers, and how and where that food is produced. A revolution in terms of environmental stewardship and reducing the carbon footprint in the food system. And finally, dignified job creation and wealth creation rooted in social justice and environmental sustainability.

The local model is economically viable. When farmers sell directly to local shops and restaurants or through farmers markets or subscriptions ("boxes"), they earn as much as 80–90 percent of the price of food. Anuradha Mittal has calculated the impact on California's economy as one example: "If just 10 percent ($85 per person per year) of Californians' food expenditures were directed toward food produced within the state, an estimated $848 million in additional income would flow to the state's farmers, $1.38 billion would be injected into California's overall economy, $188 million in tax revenue would be generated, and 5,565 jobs would be created."[42]

The political and economic ramifications of localized food systems operating in harmony with nature can be as lucrative as they are beautiful.

WASTE

We humans—especially we Americans—are literally trashing our environment. There's a great twenty-minute film by my friend Annie Leonard, called *The Story of Stuff,* that lays out the whole system by which goods are produced, starting with the extraction of natural resources from the environment and moving along to the factories where stuff is made, to the retailers who move stuff as fast as possible, to us consumers, to the dump. Along this continuum, we're producing mountains and boatloads of trash. Astronauts have looked down on dumps that are visible from space and rival the Great Wall of China in size.[43] There's a 3.5 million ton heap of debris called the Great Pacific Garbage Patch that's twice the size of Texas floating in the Pacific Ocean.[44] Houston, we have a problem.

The average American throws away four and a half pounds of garbage daily. And for every garbage bin of waste an individual puts out, seventy garbage bins come from each factory that makes stuff during the production process.[45] Dumps—and especially incinerators—pollute the air, land, water, and our bodies with toxic chemicals like dioxins. It's an altogether unsustainable situation.[46]

A handful of states and municipalities have adopted zero-waste plans, setting targets for waste reduction in the meantime.[47] Massachusetts, for example, aims to reduce municipal solid waste by 70 percent by 2010. The city of Seattle is recycling 60 percent of all the waste it generates in 2008.[48] Even so, we still have a lot of trash to deal with. And dealing with it in a smarter way could generate hundreds of thousands of jobs.

Unfortunately there's been a persistent (and understandable) stigma in many low-income communities of color about working with waste. There was a time when the occupation of sanitation worker (a.k.a. trash man) was virtually a separate caste, relegated to black or brown men who were treated and paid badly. It's only recently that the dirty, toxic operations of waste management have been complemented by the more uplifting work of recycling and reuse, with its decent, purposeful jobs. If we changed the job title to "recycling technician" and improved the pay, work in material reuse and recycling could be another sector of growth for green-collar jobs.

"In the U.S., on a per-ton basis, sorting and processing recyclables alone sustains eleven times more jobs than incineration does," says Leonard. More than 56,000 recycling and reuse centers across the country already employ over 1.1 million people. Entry-level jobs in recycling include collection, sorting, driving, and loading and can lead to advanced positions like operations manager. Entry-level jobs in materials-reuse operations include salvaging, sorting, driving, warehousing, packaging, and retail sales, after which an employee can move up to warehouse manager or floor manager.

Growth in the sector is being encouraged by measures like California's 1989 Integrated Waste Management Act, which required cities to divert 50 percent of their solid waste from landfills by 2000. California also designated forty Recycling Market Development Zones and provision of low-interest loans up to $1 million for businesses using recycled materials. As a result, in its first eighteen months, the Oakland/Berkeley Zone generated $8.2 million in investment for recycling, creating 155 new jobs and diverting 100,000 tons of new material from landfills.[49]

Specialized types of materials reclamation and reuse are cropping up. Two of the significant areas producing green-collar jobs are computer recycling and deconstruction of building materials.

Computers

At Chicago's Household Chemical and Computer Recycling Facility, students participating in the city's Greencorps job-training program learn warehousing skills; how to refurbish computers for use in schools, community centers, and low-income households; and how to disassemble the computer for individual component-part recycling. A nonprofit called Computers for Schools partners with the city to provide training. They've placed more than twenty-five thousand computers since getting started in 1991.[50]

The Alameda County Computer Resource Center in Berkeley likewise recycles computers—as well as VCRs, televisions, and copy machines. It provides on-the-job training in repairing computers, identifying and extricating computer components, and installing and using computer software. Local rehabilitation programs, homeless shelters, and parole officers refer potential employees, including homeless and mentally ill individuals as well as others with barriers to employment.[51]

Building Materials

The Environmental Protection Agency (EPA) estimates that 136 million tons of building-related construction and demolition waste are generated every year. The debris consists of a wide array of materials including wood, concrete, steel, brick, and gypsum, making for a complex waste stream.[52] Rather than demolishing this material to useless bits, buildings can be *deconstructed*, a systematic process by which valuable materials are recovered in usable form. This conserves landfill space, reduces the need for new materials, and can reduce overall building-project expenses through reduced purchase and disposal costs. The deconstruction process also keeps materials local, reducing the need to transport materials, and provides solid, family-sustaining jobs.

A nonprofit in Baltimore called Second Chance launched its architectural salvage and deconstruction services in 2003. Over the next four years the company grew quickly, filling a 120,000-square-foot warehouse space and engaging more than fifty employees— three deconstruction crews and a retail store crew. The crews consist entirely of local low-income residents who are trained on the job. Second Chance's founder, Mark Foster, explains: "Deconstruction is time-consuming and exacting. The architectural elements must be removed from the building without becoming damaged. Elements that are too large to remove intact must be removed in pieces to be reassembled later."

Foster established contracts with the City of Baltimore that call for workforce-development funds for training and first dibs on government buildings scheduled for takedown. Trainees, who are recruited through the City's workforce-development programs, receive sixteen weeks of training. The program covers a range of carpentry and craftsmanship skills such as sandblasting, painting, and stained-glass and wood repair. Once the training is completed satisfactorily, the worker is guaranteed a permanent job with the company, making between $12 and $25 an hour plus benefits.

Second Chance trainee Durrell Majette says: "As I ride down the streets now, I find myself looking at doors, the way the windows are built, the frames, stuff I used to never even notice. It's like I have a new direction in new life. I feel like I'm part of something bigger."

Foster is running trainings in Philadelphia and Washington, D.C., and hopes to open retail stores like Baltimore's in both of those cities. "We are not just offering a good job, but employment in a growing company and sector of the economy. That's a pathway to a career."[53]

Meanwhile, in the South Bronx, Green Worker Cooperatives might respond: "Employment is great, but ownership is better." Omar Freilla had the idea to found a local green co-op after serving on a coalition of labor and community groups that aimed to get green businesses to relocate to the brownfields that litter many low-income communities of color like the South Bronx.

"But I was never really comfortable with that approach," admits Freilla. "The way I saw it, if history is any guide, the profits generated by businesses—even green businesses—always leave the workers' community and largely enrich the predominantly white, middle-class and upper-middle-class owners. They live somewhere else." Surrounded by dirty waste–transfer and toxic waste–processing facilities in his native South Bronx, Freilla had an epiphany: What if we didn't throw it all away? What if we could sell it instead?

Soon after, his incubator launched its first worker co-op: ReBuilders Source, an 18,000-square-foot retail warehouse for surplus and salvaged building materials recovered from construction and demolition jobs. Among the items for sale are stainless-steel sinks, porcelain toilets, cans of paint, and doors—many of them brand-new, all of them in good shape and available at a significant discount.

There are four worker-owners, Julie Falu Garcia, Yasin York, Gloria Walker, and Carlos Angel, who manage all aspects of the business from soliciting and removing materials to warehousing and selling them. Each of them is making $35,000 a year, a good deal more than the median estimated household income in the district, which was $21,100 in 2006.

Freilla recruited local South Bronx residents by posting flyers around the neighborhood that said "Fire Your Boss" and offering the Co-op Academy. In the evening, once a week for eight weeks, the Academy provided workshops that explained and explored what a worker co-op is and what environmental justice is, along with tips on how to facilitate meetings or run consensus-based decision making. The Academy is becoming a regular event in the community, with the intent of nurturing further cooperative ventures.

"Not only do co-ops have the benefit of keeping wealth in the community; they also make democratic decision making a part of daily life. And true democracy is something that most people don't experience," Freilla says proudly. "We can set up our society so we no longer produce anything that needs to be thrown away, and we don't throw anything away, but keep it in circulation. And when you do that, you reduce waste, create new jobs, and preserve natural resources."[54]

WATER

On this blue planet of ours, freshwater scarcity is an issue of increasing concern. Retooling our society to be smarter about water can help cities save money—and also generate new jobs.

But first, some disturbing facts. Of the Earth's water, just 2.5 percent is fresh, and most of that is ice or snow. Unfrozen, liquid freshwater is mainly found underground as groundwater.[55] Unfortunately, we've covered most of our cities with nonporous materials like concrete. Therefore, a lot of rainfall runs off and becomes "storm water" instead of replenishing the underground reserves as nature intended.[56] As a result, in many of our cities, we are using groundwater faster than the rain can refill the aquifers.

With our growing human population needing to be fed by thirsty crops, freshwater supplies are shrinking fast. Climate change is also taking its toll, melting precious frozen supplies of freshwater into the salty seas and causing storms and floods of ever increasing intensity. Those heavy rains in turn magnify our storm-water problems.

Urban forestation is a key solution to the problem of storm water. This means planting and maintaining trees, yes, but also installing green roofs, which are covered in vegetation that is planted on top of a waterproof membrane. Since 2000, the green-roof phenomenon has been flourishing in cities—especially in Europe and, newly, Chicago—to absorb water that would otherwise run off into storm drains, with the added benefits of absorbing carbon dioxide, cleaning local air, cooling the building underneath (thus lowering its energy consumption), and extending roof life.

Sustainable South Bronx (SSBx), the organization founded by my colleague the MacArthur "Genius" Award–winning Majora Carter, champions urban vegetation as one solution for the pollution-ravaged South Bronx community. It cites not only the above advantages, but also the improvement of public health and building of pride in the neighborhood. Sustainable South Bronx has developed a training and placement program called Bronx Environmental Stewardship Training (B.E.S.T.).

Carter explains: "Nearly all B.E.S.T. students were previously on public assistance, and about half have prison records. They range in age from about eighteen to forty-five, and learn landscaping, green-roof installation, and brownfield remediation. After four years, 85 percent of our graduates have good, steady jobs."[57]

As B.E.S.T. program graduate James Wells explains with regard to his job planting and maintaining trees and green roofs in the South Bronx community: "It's not just transforming the environment, it's also transforming attitudes. If people see someone cares, it gives them hope, and now they care, and change their behavior. When I first started there was a lot of trash and debris, but now residents and business owners assist us. That's the best part."[58]

There's more good news. Just as in the energy sector, conservation and efficiency measures can increase our supply of water. In *Blue Gold: The Fight to Stop the Corporate Theft of the World's Water,* Maude Barlow and Tony Clarke write: "With technologies known and available today, agriculture could cut its demands by up to 50 percent, industries by up to 90 percent, and cities by one-third, with no sacrifice of economic output or quality of life." For starters, gray water (wastewater produced in domestic processes like showering) can often be used in place of clean water. We also need to widely embrace waterless, odorless composting toilets, sophisticated models of which are already available.[59]

In the meantime, there is work to be done: installing low-flow toilets, repairing leaky pipes, and replacing inefficient irrigation with high-efficiency sprinklers and drip irrigation. A regulation requiring all new residential toilets sold since 1994 to be high-efficiency and low-flow has successfully reduced the water used in flushing by 70 percent in U.S. cities.[60] There are jobs for qualified plumbers in installing toilets and repairing leaky pipes, although not necessarily for entry-level workers.

And more jobs are on the way. In the area of urban water planning, the most exciting work in the country is being spearheaded by a man in Los Angeles named Andy Lipkis. His organization, TreePeople, has planted millions of trees and taught thousands of young people the value of trees, recycling, and freshwater.

Because two-thirds of the city is paved, Lipkis explains, all the rainwater rushes into storm drains, picking up gunk from streets and sidewalks, and moves quickly to the concrete-lined LA river and the beach. Not only does the city lose all that valuable water; it has to pay fines for spewing polluted water into the ocean. But Lipkis saw a remarkable opportunity within this skewed system— to work *with* nature, rather than against it, and save the city a lot of money in the process. His idea was to build cisterns to capture the rainwater—and create jobs doing it.

The concept of building structures to catch the rain is an ancient one. TreePeople and its city and county partners have installed six demonstration cisterns in Los Angeles, which on average collect 1.25 million gallons of water for every inch of rainfall. When paired with redirected downspouts, low walls of soil known as berms, and porous ground (which comes from ripping up the concrete and re- placing it with soil and groundcover), virtually none of the precious freshwater falling from the sky is lost.

Officials were so impressed with TreePeople's projects that the city and county gave the green light to build a $200 million cistern project in Sun Valley, a flood-ravaged community mostly populated by low-wage workers. TreePeople predicts that about three hundred jobs will be created, including manufacturing and installation of water-capture systems, adapting landscaping to function as water- shed, and maintaining the landscapes, trees, berms, cisterns, and other elements of the system.

Better still, some city sanitation workers will be transformed into watershed managers. "A watershed manager is a water manager, a waste manager, an energy and resource manager—a job that takes more time and has more dignity," he points out. "The old way of dealing with water—erecting treatment plants or allowing rain- water to escape into drains, for example—actually eliminated jobs after the initial construction was finished. But watershed management creates all kinds of jobs."

Over a period of thirty years, this approach will save city and county an estimated $300 million in water and other costs. That's more than enough to pay for the retrofits, the installations, and the system's ongoing maintenance, which is where a lot of jobs are created.

Lipkis's efforts in LA have been fifteen intense years in the making, but he predicts that similar projects in cities across the country will start moving more quickly as water scarcity hits home. If he is right, many green-collar jobs will have a tinge of aqua blue.[61]

TRANSPORT

As the price of gas continues to skyrocket, we will need to reimagine, redesign, and rebuild our transportation systems and infrastructure. A massive investment in public transportation would immediately help the poor, create long-term jobs, and cut greenhouse gases.

Even with soaring fuel costs, Americans are still in love with their cars. Cars, trucks, and airplanes account for roughly two-thirds of the petroleum we consume. Each gallon of gasoline burned pumps twenty-eight pounds of carbon dioxide into the atmosphere—nineteen from the tailpipe and nine from upstream refining, transporting, and refueling.[62]

The federal government is not helping. "Four times as much federal money goes into roads as goes into transit projects," says Sam Zimmerman-Bergman, project director at the Oakland-based technical assistance group Reconnecting America and the Center for Transit-Oriented Development.[63] As a result, fewer than 3 percent of trips are made by public transit. If we increased that number to 10 percent of all trips (about the European level), we could reduce our dependence on oil by more than 40 percent, which is nearly as much oil as we import from Saudi Arabia every year.

It must be said that our failure to fund public transportation really hurts low-income Americans. The poorest fifth of Americans spend 42 percent of their annual household budget on the purchase, operation, and maintenance of their cars. That's more than twice the national average. Low-income people typically have older cars and more unexpected repair costs. More than 90 percent of former welfare recipients do not have access to a car, and yet three in every five jobs suitable for welfare-to-work program participants are not accessible by public transportation. Better bus service alone would free low-income people from car-related expenses—and expand the number of workplaces they could get to.[64]

Also, big public transit projects could be a tremendous source of jobs. Already, public transportation is a $44 billion industry that employs more than 360,000 people. More than 50 percent of these employees are operators or conductors. Thousands of others are employed in related services such as manufacturing, construction, and retail.[65]

There is no reason not to make this shift. Jobs lost in Detroit's auto industry would be more than offset by new jobs in manufacturing equipment, installing or building transit infrastructure, and maintenance and service for mass transit. Economic studies prove that transit investments actually create many more jobs than highway construction boondoggles: per $1 billion invested only 42,000 jobs are created in highway construction versus 80,000 in transit capital projects and an additional 100,000 jobs in transit operations. And many more local and long-term jobs are created in transit than in highway construction.[66]

If you want proof that communities can increase good jobs while pushing for better public transportation, look no further than the Alameda Corridor Jobs Coalition. It set an important precedent recently through its involvement in a $2.2 billion major rapid-rail project in Los Angeles. By convincing the Transportation Authority to make sure that local residents got at least 30 percent of the work hours, the coalition was able to steer a projected $40–$50 million in wages to local people. The coalition also got the Transportation Authority to require the project's prime contractor to pay for a thousand paid preapprentice training slots, so that local residents would not be excluded for lack of skills.[67]

Other countries have much smarter systems for moving people around. Picture a city of 1.8 million people with a spiderweb of bus lines connecting people to literally every part of town. Each triple-compartment bus can hold up to three hundred passengers. Passengers pay one low price to get anywhere in the city, and pay at the entrance to the bus stops rather than on board. This makes for faster loading and less idling and air pollution. Many of the city's major thoroughfares are reserved for pedestrians only, which draws more community activity into the center and more business to shopkeepers. This is no pie-in-the-sky scenario—it's been up and running in the Brazilian city of Curitiba for several *decades*.[68]

If we want to make a dent in the dual crisis, we will follow that city's wise example in every significant metropolis in the United States. If we want to employ a lot of people, help the poor, and cut back on greenhouse gases, we should cut back on building highways and cars—and invest massively in constructing buses, light-rail cars, and mass-transit projects.

ALL OF THE efforts we just examined are remarkable, in large part because most are moving upward against the force of gravity.

The old rules, regulations, laws, and subsidy schemes of the dying, pollution-based economy are still holding them back. One can only imagine what innovators of projects like these will do, once they have the government on their side.

Fortunately, help is on the way. For instance, in December 2007, President George W. Bush signed the Green Jobs Act as a part of the 2007 energy bill, to train workers for green-collar jobs. The bill was put forward by Hilda Solis (D-CA) and John Tierney (D-MA) and strongly supported by Speaker Nancy Pelosi (D-CA). It authorizes $125 million for green-collar workforce training programs. Twenty percent of those funds are targeted for veterans, displaced workers, at-risk youth, and families in dire poverty.[69] Those relatively modest funds constitute a small down payment on the massive investments that the Green Growth Alliance must someday soon require of the government, but they represent an important step in the right direction.

In the next chapter, we will discuss the full set of specific policies that will be needed to aid the pioneers discussed above—and to accelerate the process by which millions of others can join them.

SIX

The Government Question

THE PREVIOUS SECTION demonstrates—and celebrates—some great news. Even in tough places never associated with hybrid cars or organic cuisine, an inspiring transition is already under way. From the South Bronx to South Dakota to South Central LA, the tender shoots of a new economy are pushing up through the cracks in the asphalt. The reality, however, remains sobering. Encouraging and instructive as they are, these early signs of hope simply are not succeeding at the scale necessary to secure the future for vulnerable communities—or for the Earth itself.

Government policies can and must play a key role in creating an inclusive, green economy—by setting standards, spurring innovation, realigning existing investments, and making new investments. Government action can ensure that we make the transition rapidly, while protecting and benefiting our most vulnerable populations.

Governments can accelerate equitable green growth in three ways. They can *regulate conduct*—setting the rules of the game, laying down the law, establishing standards, and telling members of society what they must or must not do. They can *invest money*—from direct spending, to offering incentives, to underwriting risk. And they can *convene leaders*—spurring the formation of new collaborative institutions that solve problems by bringing together public, private, and nonprofit stakeholders.

All levels of government—federal, state, and local—will be needed to remove the barriers to green growth. And if you doubt the U.S. government can use its power to help solve big problems, I offer some reminders from the nation's history.

In 2008, old-timers across the country gathered for reunions on the occasion of the seventy-fifth anniversary of the Civilian Conservation Corps (CCC), which FDR created to alleviate the predicament of millions of unemployed people (predominantly white men, it's true) during the Great Depression.[1] When the scope of the economic crisis was understood, President Roosevelt wasted no time. He called Congress into an emergency session to authorize the program virtually overnight. The Departments of Agriculture and the Interior were responsible for planning and organizing work to be performed in every state of the union. The Department of Labor, through its state and local relief offices, was responsible for the selection and enrollment of applicants.[2]

With the CCC, the government engaged more than two million men in protecting—and sometimes developing—America's natural resources. In return for their labor, the corpsmen received monthly checks of $30 from Uncle Sam as well as basic housing and meals. The CCC is estimated to have completed projects that included planting 2–3 billion trees, controlling erosion on 40 million acres of farmland, and providing mosquito control on over 240,000 acres of land.[3]

In launching the program, FDR announced: "It is my belief that what is being accomplished will conserve our natural resources, create future national wealth and prove of moral and spiritual value, not only to those of you who are taking part, but to the rest of the country as well."[4] First-person accounts of corpsmen from that time reveal they didn't value only the money, the regular meals, and the basic literacy skills they received. They also appreciated the less tangible benefits:

> After everything I'd been through, I'd thought I was an adult when I started with the program, but it takes a tempered-steel file to put the edge on an axe. I mean that in two ways. For one, I actually learned how to sharpen an axe and a cross-cut saw. In the second way, I had the edge put on my manhood by learning how to live with others whose background was not mine, the only thing we had in common being our youth. . . . I learned a lot of skills I still use, but more than anything, it gave me my confidence back, made me feel like I was a worthwhile member of our great country, and for that I'll always be grateful.[5]

Soon after FDR's New Deal, our government decided to intervene on a global scale during World War II. A massive war effort was undertaken: a draft of men into armed forces, of women into war-related industries—remember Rosie the Riveter—and of resources into the vehicles, meals, uniforms, and so forth needed by our armed forces. The government raised taxes and strongly encouraged citizens to purchase war bonds, using posters and films and teaming up with Hollywood celebrities to amass popular support. It rationed food and fuel. And—amazingly—it even froze free enterprise, decreeing that business and labor must be subservient to the country's security interests. As one example, FDR told automobile manufacturers that they couldn't make or market any new passenger cars; instead, the country needed them to use their factories to produce the trucks and the tanks the nation needed to defeat Hitler. Detroit turned on a dime, producing at lightning speed the great arsenal for democracy that saved the world from Fascism.

And less than twenty years after that, President John F. Kennedy boldly announced the Apollo Project, to put an American man on the moon (and bring him safely back) within the decade. The president got virtually every penny he requested for the lofty project on the grounds that it contributed to national security. But the project also had widespread impacts on scientific and technological developments—arguably more important in the long run. And, of course, it made a tremendous impact on our national pride. As JFK noted during a speech several months into the Apollo Project: "[It] has already created a great number of new companies, and tens of thousands of new jobs. Space and related industries are generating new demands in investment and skilled personnel."[6]

My friends at the Apollo Alliance adopted JFK's original Apollo Project as their inspiration (and leitmotif) in their mandate for a clean-energy revolution. As the Apollo Alliance's former executive director Bracken Hendricks writes in his book *Apollo's Fire* (coauthored with Congressman Jay Inslee), JFK's Apollo Project "proved the importance of backing vision with policy and investment. Meeting the challenge meant making a commitment to expanding the capabilities of the nation in both industrial might and intellectual prowess."[7]

In other words, there are precedents—and many more than the three I've noted here—for government support of paradigm shifts and massive world-changing projects. And so, today, the U.S. government must again make a fundamental shift. Right now, the government is spending tens of billions of dollars supporting the problem makers in the U.S. economy—the polluters, despoilers, incarcerators, and warmongers. The time has come for the nation to give greater support to the problem solvers—the clean-energy producers, green builders, eco-entrepreneurs, community educators, green-collar workers, and green consumers.

TOP PRIORITIES FOR THE NEW ADMINISTRATION

The Bush administration has been a disastrous failure in the areas of environmental stewardship, climate leadership, economic renewal, and a sane energy policy. At this point, it is tempting to say that we don't need a U.S. president who will fix everything; we just need one who will stop breaking everything. That alone would make a tremendous difference. But to ensure the survival and success of our society, the president must take a number of positive steps.

First, the new administration would be wise to fully embrace the agenda of the climate-solutions group 1Sky. That organization has fashioned an ambitious set of goals based strictly on what the world's scientists say is minimally necessary to avert a global climate catastrophe.

Following 1Sky's lead, the new administration should vow to enact policies that will: (1) create five million green jobs as a part of a plan to conserve 20 percent of our energy by 2015; (2) freeze climate pollution levels now, then cut them to at least 25 percent below 1990 levels by 2020 and 80 percent by 2050; and (3) ban the construction of new coal plants that emit global-warming pollution, promoting renewable energy instead. Better yet, the president should publicly pledge to meet Al Gore's challenge of making the United States 100 percent free of fossil fuels by 2018. Such bold proposals would immediately signal the end of the status quo, stun the pro-pollution contingent, and begin to rally the nation to meet our crises head-on.

To begin making good on those commitments, the administration would then need to implement multiple policies aggressively, immediately, and at various levels. The task of meeting these challenges would do more than determine the administration's environmental policy. It would also shape America's core economic program, foreign policy agenda, urban and rural policy, and manufacturing agendas as well.

With Bracken Hendricks of the Center for American Progress, I have outlined three policy tracks the new administration must pursue simultaneously to make dramatic, politically sustainable progress on the climate, energy, and jobs policy. The first track involves exerting immediate leadership within the executive branch, taking measures to coordinate U.S. climate and energy policy across all federal agencies and using executive orders, public communications, and other presidential prerogatives to manage carbon, capture energy savings, and promote renewable technologies.

Second, the White House must engage Congress to pass a suite of global-warming and energy legislation, including both a cap-and-trade bill that limits emissions and complementary policies that strengthen standards and drive investment in clean energy. The third track will entail a vigorous diplomatic effort to reclaim U.S. moral leadership abroad through progress on international climate negotiations, clean development, and addressing adaptation and energy poverty.

1. Executive Branch Leadership

The new administration will have many opportunities for executive leadership and agency action in moving the nation onto a sustainable energy path. The president should take bold and immediate action.

Use climate solutions to frame a positive domestic economic agenda. The president will need to elevate global warming and energy security to the status of a major national commitment. He must place the issue at the center of his agenda for economic opportunity and reconstruction and link it to job creation, rebuilding cities and rural economies, and restoring global competitiveness. This will galvanize new constituencies for action, including labor, business, urban, farm, civil rights, and other stakeholders. By clearly communicating the economic benefits of action for the poor and middle class and for ratepayers and small businesses, the new administration will be able to answer predictable attacks as businesses and markets adjust.

Use the pulpit of the presidency to signal serious commitment. The power of the Oval Office to convene industry and interest groups and drive a national consensus for action should not be underestimated. Efforts should include strong signals in the opening days of the administration, including major national summits and prominent public addresses like the inaugural and state of the union, to underscore the centrality of this issue in defining the leadership and legacy of this administration, aligned with the future not the past.

Build a leadership structure within the White House to sustain this focus. To assure that attention is sustained through the hard political fights and coalition building that will be required, the new administration must establish a key presidential adviser supported by a strong office for building and implementing global-warming strategy and approaches to building a green economy. This leadership and staffing structure should be publicly launched in the early days of the administration and given authority to report directly to the president. It should have strong links to economic and national security advisers and clear pathways of communication with all agencies and White House offices to ensure that a unified strategy is employed across the executive branch.

Enlist all federal agencies in building climate solutions. Agencies across the federal government must play a role in solving the climate crisis. The administration's energy and climate strategy must be systematic and include all line agency budgets and programs.

The Department of Housing and Urban Development can advance community development and housing retrofits. The Department of Labor must ensure that a trained green-collar workforce is available. Agriculture has authority related to biofuels and wind energy. The State Department must play a central role in jump-starting international negotiations, and the U.S. Agency for International Development should shape assistance to impacted countries. The Department of Transportation should guide strategies for expansion of rail and transit, land-use planning, reducing vehicle miles traveled, and air quality and congestion.

The Department of Energy, Environmental Protection Agency, National Oceanic and Atmospheric Administration, Department of the Interior, and others will all play leading roles in the policy and science of climate change, and global warming will increasingly organize their work. Regulatory agencies like the Federal Regulation and Oversight of Energy will shape rules and incentives for smart-grid infrastructure. Meanwhile, Treasury and Commerce will establish mechanisms for carbon trading as well as incentives and financing for public infrastructure, efficiency retrofits, renewables, and transitioning U.S. industrial production.

Utilize the power of executive orders and presidential leadership. Executive orders can play a useful role in immediately implementing policies and using federal powers to make carbon-emission reduction a top priority. The White House could instruct agencies that greenhouse-gas emissions should be analyzed to achieve compliance with the National Environmental Performance Act. It could immediately grant waivers under the Clean Air Act to begin regulating carbon dioxide as a pollutant in automobile tailpipe emissions, a measure that the Bush administration denied in 2007.

Ensure that the federal government leads the way to economic transformation. Federal agencies can do much to accelerate the transition to a clean-energy economy. The president can show immediate leadership by instructing agencies to reduce their carbon footprint through improved purchasing and acquisitions, vehicle fleet management, and facilities management. Such administrative changes would set a tone of urgency and leadership that starts at the top of government.

Launch a signature initiative, the Clean Energy Corps. A national Clean Energy Corps would combine service, training, and employment efforts, with a special focus on cities and neglected rural communities, to combat climate disruption. The work would focus on retrofitting homes, small businesses, schoolhouses, and public buildings; preserving and enlarging green public spaces; applying distributed renewable energy production technology to underserved communities; strengthening community defenses against climate disruption; upgrading infrastructure; and educating children and communities on how they can contribute to ending global warming. These efforts could pay for themselves through energy savings, making the CEC program largely self-financing, while generating enormous demand for new jobs in communities that need them.[8]

A related program, the Civic Justice Corps, could provide the vast numbers of people returning from prison with a path to living-wage green jobs and careers. An excellent training model has been developed by San Francisco State University professor Raquel Pinderhughes. First implemented in Oakland, California, under the leadership of Ian Kim, director of the Green Collar Jobs Campaign at the Ella Baker Center, and also being piloted in Cleveland and Philadelphia, the Pinderhughes Green Job Corps Training combines classroom and on-the-job training, wraparound support, and an internship component. A key innovation of the model is the creation and integration of local Green Business Councils, with the intent of organizing potential employers and providing them with incentives to hire corps members.[9]

2. A Comprehensive Legislative Agenda

Cap, collect, and invest. The highest priority for the new administration is to work with Congress to pass major global-warming legislation that reduces greenhouse-gas emissions.[10] One indispensable component of cap-and-trade policy is the auction of a substantial portion of emissions permits available to greenhouse-gas emitters. The Congressional Budget Office estimates that the monetary value of these permits would range from $50 billion to $300 billion each and every year (in 2007 dollars) by 2020.[11] This money can be invested in the public interest—to equitably transition the country to a low-carbon economy.

Establish the clean-energy smart grid. The second biggest priority for the new administration—besides getting a price put on carbon emissions—must be the construction of a national smart grid for energy. The nation's present energy infrastructure is an outmoded and inefficient patchwork. A wide multitude of renewable sources cannot easily plug into it, and neither the generators nor the purchasers of power ever get enough consistent information to make intelligent decisions. Smart-grid proponents suggest another, better way: the digital automation of the entire energy supply, from the generators to the consumers. Think of the smart grid as an Internet for energy. The grid would be comprised of a network of smart devices, all communicating with each other, to do real-time balancing of energy need and production. As waste is greatly reduced, carbon reduction and cost savings would follow. In the meantime, the project would create tens of thousands of jobs for everyone from electricians to computer programmers.

Whereas our present grid is built around the needs of big, polluting, centralized power plants, a smart grid can be designed to easily accommodate multiple power producers. That means thousands of home producers of energy, perhaps deploying wind and solar, could plug into the power grid to sell or share their energy with others. Once technology for hyperconductive power lines is perfected, a national grid could even move clean energy from windy and sunny places to those regions that lack abundant supplies of renewable energy. Overhauling, linking, and digitizing the U.S. power grid is the biggest and most important piece of work in helping to move the country into a rational, clean-energy future.

There is an added bonus to this idea. A national network of more independent sources of power—home-based fuel cells, stand-alone solar systems, regional wind farms—would make the energy infrastructure as a whole a less inviting terrorist target and much more resilient in the face of natural or human-made disasters. To the best of our knowledge, no terrorist cell has ever blown up a wind turbine.[12]

Support green jobs and worker training. Improving energy efficiency and deploying renewable technology at scale will produce massive demand for skilled labor. Investing in worker training, supportive employment services, manufacturing extension, and community development will be essential to ensure we meet our goals. Legislation like the Green Jobs Act and the Energy Efficiency and Conservation Block Grant program offers an opportunity to use public investment to prime new industries as well as to lift people out of poverty. This legislation not only connects to people's immediate self-interest, but also calls them to the larger moral purpose of cocreating solutions. It is grounded in neighborhood-level actions—restoring communities with green space and green buildings, restoring bodies with parks and clean air, restoring families with purpose and paychecks.

Improve efficiency in energy generation, transmission, and consumption. To achieve immediate efficiency gains, the new administration should implement a National Energy Efficiency Resource Standard to require utilities to cut energy use 10 percent by 2020. Current building stock is wasteful and inefficient—each nonweatherized building is an open spigot for pollution and wasted energy dollars. The administration should work with Congress to pass a range of efficiency policies, including commercial and residential building codes, retrofitting public buildings to higher standards; establishing incentives for distributed energy; extending energy-efficient home mortgages; and assisting low-income and public housing stock to improve energy efficiency through stronger incentives, better accounting tools, and loan guarantees. Jobs weatherizing U.S. buildings *cannot* be outsourced.

Increase production of renewable electricity. America needs to fully deploy its abundant renewable energy resources, including wind, solar, biomass, sustainable hydroelectric, geothermal, and wave/tidal. At a minimum, the administration should require that 25 percent of our electricity comes from renewable energy by 2025. As the market for renewables grows (with technological improvements and economies of scale), the objective should be to drive the price cheaper than traditional fossil-based energy in the market, allowing a sunset on any financial incentives. Diversifying electricity and fuel supplies hedges against disruptive spikes in energy costs. Renewable electricity creates *more than twice as many jobs* per unit of energy and per dollar invested than traditional fossil fuel–based electricity. And electricity and heat account for more than 30 percent of all U.S. carbon emissions, a figure that can be drastically reduced by turning to low-carbon, renewable energy.

Invest in low-carbon mass transportation and rail infrastructure. In an era of escalating oil prices, traffic gridlock, hazardous air quality, and the threat of a global-warming tipping point, the president must reinvest in local mass-transit systems, regional and interstate high-speed rail, and other low-carbon means of transportation of both passengers and freight. Expanding mass transit and rail infrastructure promises to create thousands of good construction jobs,

while expanding Americans' transportation choices and strengthening communities.

Increase vehicle fuel economy. In the early years of cap-and-trade,
the price for carbon will probably be too low to change driving behavior. A price of $15 per ton of carbon dioxide translates roughly
to an increase of 13 cents per gallon in the price of gasoline, not
enough to dramatically reduce gasoline consumption. An additional increase in the Corporate Average Fuel Economy (CAFE)
standard (to at least 40 miles per gallon) will be needed to spur
vehicle efficiency. Increased CAFE standards, combined with incentives for auto manufacturers to retool factories and for consumers to
purchase more efficient and alternative fuel vehicles, will support a
resurgence of American automotive manufacturing and help move
us off both carbon and oil. A new line of ultraefficient vehicles, such
as plug-in hybrids that get 100 miles per gallon of gas, will support
manufacturing jobs of components—from new battery systems to
advanced drive trains—and help consumers.

Change the systems for fueling our bodies. Federal policies and
programs can move our farm and food system in a more sustainable
direction. Current policies subsidize farm consolidation at the expense of family farms, reduce opportunities for new folks interested
in pursuing farming, and encourage industrial, pesticide-based
farming practices that hurt the health of the land and consumers.
Instead, federal grants and funding could support:

- Fair labor, fair trade, and environmental standards

- A national organic-transition support program that facilitates
  the adoption of organic standards by conventional farms

- Community-based health-food and nutrition initiatives

- Micro-credit programs to encourage new entrepreneurs in
  farming and food production

- A matched savings fund for new farmers

- Crop insurance that rewards crop diversification and does not discriminate against organic farmers

- Amendment of the Packers and Stockyards Act to eliminate price preferences that discriminate against small and mid-size livestock operations

- Sustainability criteria to guide the development of agriculturally based renewable energy

- The Outreach and Assistance for Socially Disadvantaged Farmers and Ranchers Competitive Grants Program

We can't use tons of fossil fuels in our fertilizers and massive Robo-tractors just to make a handful of carrots. We need to start preparing now to transition our entire food system to one that relies less on petroleum and pesticides and more on smart people and wise farming techniques.

Block new coal plants that can't safely capture and store emissions. Next-generation coal-fired electricity faces economics similar to those of cars. The cost of carbon dioxide emissions would have to reach roughly $30 per ton before it would be economically rational to deploy advanced technology that captures emissions (carbon capture and sequestration). It may be several years, or even decades, before carbon prices reach such a price threshold—and we don't have that much time. To freeze its emissions and begin serious reductions within the next few years, the coal-fired electricity sector will require a tough performance standard. For the sake of the planet and all future generations, no new coal plants should be permitted unless and until they can safely capture 100 percent of carbon dioxide emissions.

In the meantime, the United States should not build new power plants based on old, dangerous, carbon-spewing technologies. If coal is to play a productive role in our energy mix, the government and private sectors must collaborate on a massive Manhattan Project to achieve dramatic breakthroughs in carbon-capture technology. Even then, the president should move to ban the devastating practice of "mountaintop removal." No coal company should be allowed to blow up mountains—destroying America's beauty, poisoning its rivers, and destroying rural communities—just to scrape out the coal deep inside them. It is time to break our addiction to dirty coal before our addiction breaks Appalachia.

Provide sustainable, low-carbon fuels. The administration should pursue multiple options to reduce dependence on fossil-based transportation fuels. At the very least it should commit to producing 25 percent of our liquid transportation fuels from renewable sources by 2025. The majority of renewable fuels should come from next-generation biofuels made from nonfood biomass like switchgrass, wood chips, and agricultural waste. To ensure the environmental integrity of biofuels, the administration should implement a low-carbon fuel standard to reduce life-cycle emissions 10 percent by 2020. It should also initiate a certification program with transparent sustainability labeling.

Also, a suite of policies are available—including zero-emissions mandates and consumer tax incentives—to accelerate deployment of plug-in hybrid and all-electric vehicles. To build the infrastructure to supply this energy, a "pump or plug" mandate should require that 15 percent of gas stations retrofit their facilities to deliver either E-85 ethanol, biodiesel, or dedicated electricity charging stations for plug-in vehicles in all counties where 15 percent of registered vehicles can run on alternative fuels. Further, local ownership can provide strong economic benefits to rural communities by producing sustainable bioenergy.[13]

Eliminate federal tax breaks and subsidies for oil and gas. The federal government currently allocates billions of dollars annually in tax breaks and subsidies to the oil and gas industry. With high prices, companies are making record profits and don't need government assistance. It is time to shift investment away from high-carbon energy to the clean energy necessary to power a low-carbon economy. Redirecting investment to help commercialize emerging low-carbon energy sources will help transform our economy and capture the economic and environmental benefits of clean energy.

Trade in "hoopties" for hybrids. "Hoopty" is street slang for an old, unreliable car—often a gas-guzzler. The term is popular in disadvantaged communities because such cars—ugly, noisy, and often belching fumes—can be found everywhere.

Such cars are the least fuel-efficient, least safe, and the most polluting cars on the road. As a result, low-income drivers can end up spending much more money for gas than wealthier people, who can afford to buy newer, more fuel-efficient cars—or even hybrids. Their pain eventually becomes everyone's pain because, in the aggregate, these cars drive up the price of gas for everyone. Maintaining a massive fleet of gas-guzzlers keeps the demand for fuel high—and thus adds to the upward pressure on fuel prices.

To address these problems, the government should adopt—urgently—a policy that helps low-income drivers scrap their inefficient vehicles and replace them with efficient cars. Such a policy would help people of modest means save money on gas; therefore, it would be a great instance of eco-populism in action—showing how green solutions can help ordinary people. It would also decrease the nation's oil and gas consumption, improve air quality, help beat global warming, and hasten the retirement of a whole generation of gas-guzzlers. An added benefit would be that the hard work of deconstructing and recycling the materials in all those old hoopties could be a source of good, green-collar jobs.

There is precedent for a program like this; a few states have already implemented scrappage programs designed to get the most polluting cars off the roads. They are working well. Also, the government already helps low-income households meet their home energy needs through the Low Income Home Energy Assistance Program. This program would help them meet their transportation fuel costs as well.

The question is how to do this in the most affordable manner possible. Eco-visionary Amory Lovins has two solutions. In the first option, the federal government buys a large volume of efficient cars and leases them to low-income people who qualify. According to the Center for American Progress: "The cost of insurance, gasoline, and regular maintenance could be incorporated into the leasing price in much the same way subscription car-sharing programs, like Zipcar and Flexcar, do now. Volume purchasing of insurance and gasoline would lower these transportation costs."

The second option involves helping people who currently do not qualify for new car loans obtain credit to buy efficient vehicles. The Center for American Progress reports: "The risk for this customer segment is similar to that for student loans with the additional benefit of the car serving as collateral. If the federal government guarantees reimbursement to current auto lenders for incremental defaults made to participants in the low-income scrappage program, the existing market mechanisms and financial institutions could be used for this program with little cost to the government."

In other words, just as with student loans, the government would not directly spend much money. Instead, it would use its financial clout to open the doors so that millions of people can get the bank loans they need to improve their own lives. Such a role for government is especially important in the wake of the home-mortgage meltdown. If there is anything we can do to prevent it, those millions of Americans who lost out in the housing meltdown should not be further impoverished every week at the gas pump.

In combination with a reduction of the speed limit back to 55 mph, a national "scrap and replace" program for polluting cars would save low-income drivers money, slash greenhouse gas emissions and lower gas prices for everyone. And of course the side benefit: it would result in serious cool points for the president. The youth of America would get a huge kick out of hearing the president of the United States say the word "hoopty" on national television. A cooler president could give us a much cooler planet.

3. Leadership in International Negotiations

One great tragedy of the Bush administration has been the abdication of international diplomatic and moral leadership—especially in the arenas of greenhouse-gas emissions and global-warming treaties. The new administration must reengage with the world and simultaneously rebuild American standing abroad. The following steps must be taken rapidly.

Rebuild international credibility through strong domestic action. The most important message we can send is through our own strong action to reduce domestic U.S. carbon emissions. The policies mentioned above will help to restore American credibility in international negotiations.

Reengage international negotiations. In addition, the United States must rapidly reengage international climate negotiations at several levels. This will start immediately in the new administration, with the next UN Climate Change Conference in 2009 in Copenhagen. These talks will set the framework for a successor treaty to the Kyoto Protocol.

Author and climate activist Ross Gelbspan has an ambitious plan to replace the Kyoto agreement, which is widely criticized for its inadequate emissions reduction targets and loopholes that allow for emissions trading and give credits for the use of carbon "sinks" and the provision of emission-reduction technologies, neither of which are adequately regulated. He proposes that the industrialized countries strip away the $250 billion a year that they collectively spend to subsidize fossil fuels—and use that money to support clean energy. Second, he proposes that government tax the $1.5 trillion worth of daily international currency transactions by a quarter-penny on a dollar and use the $300 billion a year generated by the tax to fund wind farms in India, fuel-cell factories in South Africa, solar assemblies in El Salvador, and vast, solar-powered hydrogen farms in the Middle East.

Last, he proposes that every country simply agree to Progressive Fossil Fuel Efficiency Standards. Every country would start at its current baseline and increase its fossil-fuel energy efficiency by 5 percent every year until the global 70 percent reduction is attained. That means a country would produce the same amount of goods as the previous year with 5 percent less carbon fuel. Alternatively, it would produce 5 percent more goods using the same amount of carbon fuel as the previous year.[14]

Connect global warming and trade policy. Climate provisions should be given significant weight in international trade policy. Industry and labor have expressed concern over adverse economic impacts from trading relationships with countries that lack controls on carbon. These concerns should not be an excuse for weak standards, but rather should drive policies that ensure a level playing field for U.S. workers in a world where carbon has a price. As domestic legislation moves forward in Congress, trade implications are likely to receive increasing political attention, and the president must have answers to calls for a border adjustment tariff, a program for trade adjustment assistance for displaced workers, and a strategy to include climate provisions and broader environmental and labor protections in future trade deals.

Promote adaptation and confront energy poverty. The secretary general of NATO has identified climate disruption as a top security challenge, resulting from water and agricultural shortages and migration of refugees.[15] These global security threats as well as moral imperatives require rapid and forceful attention to two major areas of development assistance. First, increasing the ability of poor countries to access food and water and to ensure public health and public safety must become a top priority for the administration. Second, we must be ready to help poor countries leapfrog pollution and rapidly deploy clean-energy technology. The goal of international development assistance should be to alleviate the crippling energy poverty that denies much of the world's population basic energy services, without increasing carbon emissions.

Among global activists, a critical mass is beginning to rally around the Greenhouse Development Rights (GDR) framework. It was developed by Paul Baer and Tom Athanasiou, of EcoEquity, and Sivan Kartha, of the Stockholm Environment Institute. The GDR recognizes that the desperately poor around the world have a right to develop themselves economically, even if they add slightly to carbon emissions. In other words, they have a right to bring themselves up to a dignified level of consumption. Meanwhile, it is the rich who must now bring their emissions and consumption down to a dignified level.[16]

The First One Hundred Days

Whatever the president does, he must act quickly. Diplomatic, scientific, and economic timetables are all running out. The president must hit the ground running. On energy and climate policy, it is critical that significant efforts be undertaken in the early days of the administration. A hundred-day strategy is key to making real domestic commitments and advancing stalled international climate talks that will result in meaningful global reductions.

The full power of the presidency, both its political leadership and moral authority, is needed to build deep public support to sustain smart climate policies over a generation. Climate solutions must be at the center of the agenda for the entire administration. This effort will also require broader constituencies in support of action and new strategies for public education. Solving global warming should become a centerpiece and organizing principle for the administration's program for economic revival.

WHAT LOCAL POLICIES CAN DO

Ultimately, the federal government must play the leading and defining role, but local and state governments have important parts to play. And many already are passing laws and creating policies that are moving our society into a brighter, healthier future.

The involvement of city governments is a very good thing, because for the first time in human history, more people on Earth live in cities than outside of them. Urban settlements cover only about 2 percent of the Earth's surface, but they consume more than 75 percent of the Earth's resources and produce 75 percent of the Earth's waste (including air and water pollution).[17] Therefore, what we choose to do in our cities, here and around the world, will either sink the planet—or save it. If we want a planet that can sustain human life, we must create human cities that can sustain the planet.

A Look at Chicago

One American city is quite clearly at the forefront, leading the way toward sustainability, but it's probably not the place you'd guess. "With its strong industrial base, Chicago is perceived to be a meat and potatoes kind of town," says Chicago's chief environmental officer Sadhu Johnston, "so for it to set a green example is different than a city like San Francisco or Boulder doing so." It's true. For such groundbreaking environmental leadership to come out of "The

City of Big Shoulders" rather than the land of tofu and hot tubs is remarkable. And the example that city is setting is a powerful one for industrial centers across the nation and the world.

Okay, perhaps you aren't completely surprised. After all, in the previous chapter, you already met LaDonna Redmond, the Chicago mother who started an organic food revolution in Chicago's black community. But it turns out that she is seriously not alone in trying to green the Windy City.

Mayor Richard M. Daley got to be well known for his commitment to bringing ecological solutions to his city—even before climate change became a front-page story. Johnston says the mayor's initial impetus was economic and social: to improve the quality of life in Chicago and make the city a more competitive place to live, work, and locate a business. Then a team of top climatologists hired by the City predicted a climate like Houston's by 2095 if current emissions weren't halted. That's when Daley added environmental considerations to a list of reasons to build a clean, green economy. (In response, the City set a goal of reducing emissions 80 percent by 2050.)

Johnston stresses that the success of the Chicago model has been building the economy from multiple angles, stimulating both supply and demand. "We're using our purchasing power as a city to attract the kinds of manufacturers we want in the city. For example, we attracted a solar-panel manufacturer by promising to buy $5 million worth of its panels. Now its factory is employing 99 percent local Chicagoans, with a number of ex-offenders among them, to manufacture these panels. Meanwhile, the copper piping and the glass panes and the other parts that go in the panels are being sourced from other Chicago companies."

Chicago also funds incentives for a number of green processes and products, for example, giving away solar-thermal panels made in Chicago to homes and businesses; making green-roof grants, which award $5,000 to anyone installing one; and running geothermal and reflective-roofing grant programs. A Green Bungalow program provides matching grants to homeowners for weatherization and/or renewable energy installation. The Green Permit Program expedites building permits and pays $25,000 toward the cost of building permits for those who are building green. The City is learning it can create financial incentives—and the incentive of speedier processes—to effectively help homeowners and business owners go green.

Last year the City developed an official Green Business Strategy to help conventional companies get greener and expand their green jobs. One of the recommendations resulted in the Waste to Profit Network. It currently has about eighty participants, including utilities, breweries, coal power companies, a huge steel manufacturer, and Abbott Labs, one of the largest pharmaceutical companies. Says Johnston:

> We do a waste audit for every participating business, and an input analysis for every participating business. And then we start to make connections around materials. We've saved them hundreds of thousands of dollars and saved thousands of tons of materials from going into landfills.
>
> For example, Abbott Labs makes IV bags, which need the edges trimmed off. The plastic that gets trimmed is not recyclable, so it has to go into a landfill, and that costs them money. But we've connected them with a tiny start-up company called Curb Appeal that's also in the network, and they have a technique for making parking lot bumpers—what your wheel butts up against in a parking lot—with that discarded plastic. Curb Appeal just needed a client ready to purchase them. So the City committed to buy several hundred of these bumpers for a parking lot we're building. The network is sort of a virtual eco-industrial park, and definitely a win-win-win scenario.

In early 2008, Mayor Daley hosted a union summit that brought together leaders from twenty-four trade unions and their apprenticeship programs to talk about expansion into alternative energy, energy efficiency, and green technology. "It built off existing partnerships we have, for example with our Local 399, the operating engineers, with whom we've been partnering to train their members about green building and how to green existing buildings, and our partnership with the electrical workers union, which has had an apprenticeship program with the City in solar-panel installation," Johnston reports.[18]

Another of the mayor's central concerns has been the reentry into society of formerly incarcerated people. About twenty thousand people are released from prison every year in Chicago. The mayor asks every program in the City to give ex-offenders meaningful ways to contribute. The Department of the Environment provides a particularly beautiful solution: the Greencorps. Since 1994, Greencorps Chicago has been at the vanguard of reentry programs geared toward green opportunities, providing nine months of job training, paid internships, and job placement in an ever increasing variety of green industries.

A complex array of partnerships between government, the private sector, and nonprofit agencies is behind the highly successful model, "so the City doesn't waste any time reinventing any wheels." For example, partner organization Fuller Park Neighborhood Development Corporation manages the home weatherization component. Meanwhile, Computers for Schools trains Greencorps members to fix or disassemble and properly dispose of old computers that are dropped off at the City's Household Chemical and Computer Recycling Facility. These computers are then distributed to low-income households, community centers, or schools, thereby helping to close the digital divide.

A workforce development agency called OAI, Inc. (formerly the Office of Applied Innovations), handles recruitment by partnering with community-based agencies. It also provides certifications in many areas: cleaning up polluted brownfields; getting rid of lead, asbestos, and mold; handling hazardous material safely; responding to emergencies; and complying with the Occupational Safety and Health Act (OSHA). OAI also helps Greencorps students with job readiness (for example, how to deal with sexual harassment), interviewing, and other support services. These are skills that serve graduates well wherever they go.

Another agency called Employ America grounds students in basic financial literacy. Employ America also serves as the employer of record for program participants, providing basic medical benefits and a matched savings program. As the employer of record, Employ America gains tax benefits by employing people from "empowerment zones" (government lingo for distressed and impoverished communities). It then turns those savings around and offers Greencorps trainees a dollar-for-dollar match for the money they save.

And this is only a sample of the more than forty groups that help make the training experience a success. Even more organizations and companies are involved on the placement side, hiring Greencorps program graduates for jobs earning anywhere from $9 to $38 per hour, with most in the $11–$14 range.[19]

With the City's adoption of a Climate Action Plan and a Climate Change Job Strategy Initiative, the program will expand even further. Johnston says the City is analyzing what kind of preparation will be necessary for a workforce that can retrofit thousands of homes, replace thousands of appliances, and install thousands of solar panels, among other activities. "Our initial estimates show this will result in thousands of new jobs per year. Those are contracting, auditing, engineering, labor, insulation—on the retrofit side. And many more jobs in expanding public transit and other initiatives."

Chicago's efforts are already impacting real people's lives at the grassroots level. Take, for example, Chicagoan Jumaani Bates, who has turned his life around and is now a part of the city's green wave.

Bates is a Greencorps graduate. He spent fifteen years in a revolving-door relationship with Chicago's Department of Corrections. Now the people he knew on the streets during his drug-slinging days in Westside constantly ask him for advice, because they see him successfully maintaining a different kind of life. He says:

> I tell them to clean up and then go to Greencorps or Growing Home. With all these new green technologies and jobs in green building and green business, we need to get the word out. African Americans were last on the list in terms of the information age, and are still trying to catch up with that. I keep telling people you need to know about this green economy, whether you believe in global warming or not, because that's where the opportunities are.[20]

Today, Bates works at Wilbur Wright College, one of the City Colleges of Chicago, in the Building Environmental Technologies program. That program trains construction-industry professionals in green building and energy conservation. He's also a member of the local chapter of the U.S. Green Building Council and of a network in Chicago called Blacks in Green. Soon he plans to pursue a degree in horticulture and natural resources at the University of Illinois at Urbana-Champaign. He has a bright future—and so does his city.

LOCAL POLICY IDEAS

Mayor Daley's Chicago illustrates perfectly the powerful combination of setting standards, investing, providing incentives, and fostering innovation. Already many other U.S. mayors and governors are implementing policies that address our environmental problems as well as our social and economic crises.

Below, I highlight some of the most interesting and exciting ideas that local and state governments can embrace. There are dozens of opportunities at the local level to support the greening of America and the creation of new jobs.

1. Green Existing Buildings

Every city should commit to retrofitting its existing buildings and permitting only the construction of efficient, high-performance structures that meet top Leadership in Energy and Environmental Design (LEED) standards. A city's commitment to a comprehensive retrofit of its own buildings demonstrates forward-thinking green leadership, results in savings in energy bills and health-care costs, and creates a ton of work. As a city works hard to make its own buildings more efficient, its leaders can in good conscience require the same of all residential, commercial, and industrial buildings in that municipality.

One big barrier to greening buildings, however, is figuring out how to pay for the work. In theory, the retrofits should pay for themselves in dollars saved on energy costs. But in practice, few building owners have the cash on hand to pay up front for all the necessary audits and repairs. So buildings keep leaking energy, property owners keep leaking cash, and greenhouse gases from overtaxed power plants keep building up in the atmosphere. The key is to create mechanisms that let the cost-saving, retrofit work be done right away, with immediate gains to property owners or renters—but then be paid off easily, over time.

As suggested elsewhere, I think we should be spending much more public money on efficiency efforts. We can do that as direct grants or as low- or noninterest loans in a revolving loan fund. It's a sensible social investment. But what about right now, when the public money for comprehensive building retrofits isn't available, and city leaders must go to banks and private investors to finance their plans?

The COWS-Milwaukee model—Milwaukee Energy Efficiency, or Me2—is a good way of doing this. The Center on Wisconsin Strategy (COWS), led by Joel Rogers, invented the model while trying to figure out how workers in the gutted industrial heartland can move to the center of the clean-energy economy. The idea is to retrofit practically every building in the city to save money and put lots of people to work.

Me2 is an innovative model that allows small-property owners and even renters to use a process similar to performance contracting in order to achieve energy savings. Property owners or renters (with landlords' cooperation) receive an audit listing all conservation measures that can be paid for out of energy savings in a given period. They repay the cost of the measures via their utility bill.

Thus far, COWS has gotten buy-in from labor, business, and community civic and political figures in Milwaukee as well as on the state level. Importantly, it also acquired assurance of private financing from JPMorgan Chase once the meter-based billing scheme is implemented. Public financing has also been promised.

COWS and the University of Florida estimate that for every $1 million invested, about 13 job-years in installation/construction activities and 4 job-years in upstream manufacturing will be created. For single-family residential projects, the distribution of jobs will be 0.5 supervisor job-years, 2.5 skilled labor job-years, 4.7 semiskilled labor job-years, and 5.0 unskilled labor job-years.[21] This is an estimate that should hold true for any city. This is great news for lower-skilled people with barriers to employment, like criminal convictions.

And the really exciting thing about the Milwaukee experiment is that it will in fact be scaled—applied to every building, made a focus of citywide efforts—from the bully pulpit of the mayor to "social marketing" of the concept by community organizations, many of whose members will be getting the new work. This is the sort of beautiful, scalable solution that could be made in any American city or town.

2. Create Green Assessment Districts

Another great idea is using the well-established notion of "assessment districts," which have been proven to work very well for other purposes, for green projects. For instance, a neighborhood decides to put its utility lines underground. Residents opt in to an "assessment district," committing to pay an extra amount, maybe $10 per month, in property taxes. The utility company does the work of undergrounding the lines and is paid back from the city out of the assessment fees. Cities have the opportunity to enact this same model of financing—which is not a tax, but a voluntary, opt-in tax assessment—for efficiency retrofits as well as renewable energy projects, in both commercial and residential sectors.

Suppose residents in an area wanted to get their homes weatherized and outfitted with solar panels. Many would hesitate. For one thing, the interest rate on any individual's home-improvement loan could be discouragingly high. Second, certain homeowners might sell their houses in just a few years and move out—before they had time to make their money back on the investment through savings on energy bills. Therefore, the renewable energy and conservation services might never get purchased—and those homes would go right on using dirty energy in a wasteful manner, despite the owners' contrary desires.

But there is a way out of this dilemma. Neighbors could work with city officials to create a green assessment district. To pay for green retrofits within that district, residents who chose to participate would agree to pay a small assessment added to their local property tax bill. When a large enough number of people signed up to participate, the City could bundle up all of those promises and take them to a bank. There, it could get a big loan to cover these home improvements en masse at a much lower interest rate. Even better, the obligation to repay the loan would not attach to the temporary owner, but to the house itself, like a green lien. Every year, whoever happened to own the house at that time would simply pay back part of the loan as a part of the property taxes. As a result, the cost of the green improvement would be spread out over multiple owners. And for any particular owner, the savings on the energy bill should cancel out the yearly property assessment cost. In fact, the City of Berkeley, California, presently is experimenting with this very model.

Such an approach would create demand for clean, green technology (like solar panels) and the labor to make it happen (like the installation of said solar panels). In fact, an agreement to hire local labor could be specified up front, so the benefits stay within the community at every stage. And those neighborhoods that have the means to opt in and adopt the costs early on end up creating the opportunities for other poorer neighborhoods to green themselves at a lower cost in the future.

Nobody in the district would be forced to participate. Only those who signed up would pay the assessment—or get the benefits. But everyone who did sign up to take part would enjoy the best of all possible worlds—immediate energy savings and home improvements without incurring heavy personal debt.

3. Establish a Carbon Budget

Cities need to assign a carbon impact—a monetary value—to all of their activities and expenditures during their regular budgeting process. This would lead to true cost accounting and show the environmental impact of a municipality's activities. Using this method, a city can assess the real costs of tearing down a building and building a new one compared to the costs of refurbishing the original structure. A carbon budget is an indispensable tool for guiding a city's budget and procurement, ensuring that a city's use of taxpayer dollars supports the clean, green economy.

As in the case of retrofitting their own buildings first, cities have the opportunity to lead by example and prove their commitment to limiting emissions. Cities should quantify and assign a value to all the carbon emissions of the municipality (or county or region)—including all industrial, commercial, and residential outputs. That process is a critical first step in ratcheting down emissions overall.

4. Set Targets for Local Food, Zero Waste, and Renewable Energy

Municipal and state governments can also lead the way in promoting local food, supporting zero waste, and boosting renewable energy usage. Even a light meal might represent a ton of carbon, depending on how far each piece of food traveled to get to one's plate. The closer to the dinner table that a meal is grown, the less distance that carbon-spewing trucks must travel to ship the food to diners. Therefore, increasing the amount of locally grown food that people in a city eat is a smart way to cut global-warming pollution—and simultaneously increase the number of regional agricultural jobs. A campaign to support local food might specify that half of the food for an urban area come from within two hundred miles. This goal could revitalize farming in rural areas, otherwise threatened by economic despair and suburban sprawl. And this goal would also bring into play opportunities for rooftop gardens, urban agriculture, and the conversion of unused urban land to urban gardens and farms. All the better if the food is not just local, but organic, and doesn't rely on hormones and pesticides (poisons).

When a city makes a zero-waste commitment, it is essentially declaring that nothing bought or used within its boundaries can be thrown away. Everything must be reclaimed or used for another purpose. A city can set interim targets, such as a plan to reduce municipal solid waste by 70 percent by 2020. The higher the targets, the greater the entrepreneurial and work opportunities for local residents in the local reuse and recycling industries. A zero-waste economy reduces waste and increases work. The reason for this is simple. Imagine two factories side by side, one putting things together and the other taking things apart, so the parts can be recombined to make other products. In this model, the city would have twice as many jobs as it would if all of the material the second factory deals with was going into a landfill or an incinerator.

Another positive step for city governments to take is setting targets for the use of an increased percentage of renewable energy. Cities set these targets by creating what are called Renewable Portfolio Standards. By setting big targets for the purchasing of clean energy, cities can stimulate investment in solar, wind, and other renewable power and enjoy the growth in jobs that follows.

Additionally, a commitment to replacing the concrete of city streets and other paved areas with permeable surfaces would address the water-management problems that many cities face. Chicago is ahead of the curve in exploring permeable coatings with its Alleyway program. Innovative gray-water systems could be installed to cycle water from streets, medians, or plantings into city irrigation or fire-hydrant systems. In an increasingly water-stressed world, rainwater should be seen as a resource, not a pollutant. When cities set high targets for renewable energy, local food, waste reduction, and even rainwater management, they help the planet and spur the green economy.

5. Use Urban Planning to Create Urban Villages

Utilizing general plans and zoning laws, we need to reconfigure how urban communities are designed and how they function. In 2030, two-thirds of the world's population will be living in urban centers—up from 50 percent today.[22]

In the past there have been powerful incentives for development to sprawl outward from city centers. But sprawl's effects are horrendous. It breaks up the social fabric of society by necessitating cars for everyone, contributes to global warming and climate disaster, and destroys farmland and green space.

The solution is "transit-oriented development," which means the intentional creation of walkable, self-sustaining, mixed-use urban villages. In urban villages, people have access to housing, shops, recreational spaces, and work—or the public transit to get them to work. Cities must commit to infill development (clustering housing and businesses within city limits, not in the suburbs). Leaders in zoning and urban planning should also make a commitment to superhigh density (expanding upward, not outward).

It is strange but true, the gray and unassuming members of the planning commission of every city hold the fate of the world in their hands.

PAYING FOR IT

We need innovators, and innovators need investment. Fortunately, ideas to put muscle behind this slate of policies and programs are in abundance.

1. Create a Federal Revolving Loan Fund for Energy Efficiency

We know that retrofitting buildings produces a return on investment of between 10 and 20 percent. However, in many cases the lack of up-front capital or financing options makes it difficult or impossible to make these investments. The federal government should set up a revolving loan fund for energy-efficiency projects in residential, commercial, and industrial buildings. The fund could issue low-interest or long-term loans to businesses, hospitals, schools, local governments, and others to green their buildings and lower their energy costs.

These loans could complement other energy-efficiency financing mechanisms and help to lower transaction costs or overcome inertia in cash-strapped cities or institutions. For example, any mayor could save lots of money by retrofitting all the city-owned buildings. And a lot of energy service companies would be willing to front the money for the work and get paid back (plus a profit) with the savings from better energy performance. This practice—called an energy savings performance contract—is becoming more widely used for many large projects. But mayors of struggling urban cities or rural towns probably don't even have the resources to research options and broker a good contract. The revolving loan fund could provide them with that capacity.

A successful revolving loan fund could even generate revenue that could be invested in worker training and national service programs, like the Clean Energy Corps. If it were large enough, it could conceivably fund grants to cash-strapped cities for greening projects.

2. Boost Investment by Backing Loans and Matching State Funds

A powerful way the government can support clean and green innovation is by guaranteeing loans. Loan guarantees make capital cheaper for businesses that are experimenting and trying to scale up new methods of water conservation, energy efficiency, and renewable-energy production as well as new products like energy-saving vehicles. When the federal government guarantees the loans, banks can loan money to these businesses at lower interest rates. This also helps drive private-sector funding to the innovators. And the flourishing new businesses create lots of new jobs.

The federal government has long guaranteed loans for industries ranging from maritime interests to nuclear power. In fact, the Department of Energy recently developed a "clean energy" loan guarantee program. But the majority of the $38.5 billion program being launched in 2008 is targeted at nuclear and fossil-fuel industries. The program needs to be redesigned to support only truly renewable energy.

Here's another inventive idea. States can—and many already do—generate money through a small charge on energy use (e.g., one-tenth of a cent per kilowatt-hour, which amounts to an extra one dollar added to an average residential monthly utility bill). They then pool this money in a public-benefits fund. A number of states use some of their public-benefits fund to finance the costs of retrofitting. Energy savings then can be reinvested in the economy, creating increased growth and tax revenue as well as environmental benefits.

The federal government could provide a dollar-for-dollar match to the money that states spend on energy efficiency. Such a program would vastly speed up the retrofitting process. And that would ultimately mean massive savings on energy bills, lessened reliance on fossil fuels and thus reduced carbon emissions, better public health, and lots of green hard hats all over the country.

3. Catalyze Green Businesses with Green Strings

The key isn't always finding new dollars; it's also reallocating existing funding. State and local governments already provide billions of dollars in economic-development incentives and subsidies every year. Greg LeRoy, of Good Jobs First, has proposed that government should attach, where appropriate and reasonable, "green strings" to all of these programs.[23] If even a small portion of the recipients of these subsidies had to meet certain green standards to qualify, it would have a profound and immediate effect on the overall "greenness" of local economies. Governments can also attach job standards to the receipt of subsidies or incentives—similar to the growing number of community benefit agreements negotiated between community groups and project developers—thereby ensuring that people who most need jobs are able to get them.

FROM A LEADER IN POLLUTION TO THE LEADER IN SOLUTIONS

Our nation has always prospered when we invested in innovative leadership in technology: from rural electrification, to new transportation networks like the transcontinental railroad and interstate highway systems, to public investment in semiconductors and the telecommunications and Internet revolutions. Bold public leadership providing incentives for scientific inquiry, new technology deployment, and infrastructure have repeatedly enabled the private sector to flourish while building a growing middle class. Today, the "clean-tech" revolution and the transformation of our aging energy infrastructure are poised to become *the* next great engines for American innovation, productivity and job growth, and social-equity gains.

With smart federal and local policies, climate solutions can be a progressive force economically. Building a clean energy economy can generate literally hundreds of billions of dollars of productive new investments on a scale equal to that of the greatest periods of past American economic expansion. And globally, the best contribution that we in the United States can make as world citizens is to roll up our sleeves and get our government on the right side of the carbon-cutting, green revolution.

SEVEN

Buoyancy and Hope

I have nothing to offer but blood, toil, tears, and sweat. We have before us an ordeal of the most grievous kind. . . . You ask, what is our aim? I can answer in one word. It is victory. . . . Victory, however long and hard the road may be, for without victory there is no survival.

> —Winston Churchill,
> Prime Minister of the United Kingdom,
> 1940–45, 1951–55

I PRAY THAT this book will give hope, encouragement, and inspiration to those who are working to move our society along the path toward a green future. I believe we can get there. But I do not want to create any false sense of comfort or security about the outcome. The fact that our survival is a vibrant possibility does not mean that it is inevitable. Far from it.

Even the most hopeful ideas discussed here are based on a fundamentally pessimistic premise: that the children of all species, including our own, are gravely imperiled—unless we completely overhaul our outmoded political and economic systems. Such a transformation would constitute one of the single biggest feats in the history of world politics. As Niccolò Machiavelli put it: "There is nothing more difficult to take in hand, more perilous to conduct, or more uncertain in its success, than to take the lead in the introduction of a new order of things."[1] Even if we had skilled leaders in both major parties who were 100 percent committed to the undertaking, it still would be tough for them to produce change of the necessary scope and scale. And so, in this venture, failure is an option. We could lose this crusade, and in losing—lose everything.

No matter what happens, we will need to brace ourselves for a rough ride. Even if our efforts at ecological salvation are blessed with unprecedented success, things will get worse before they get better—probably much worse. In other words, someday we will probably look back on these tough and turbulent times and see them as the "good old days." We may someday sorely miss the kind of relative ecological, economic, and social stability that we are presently enjoying.[2]

But even if our movement fails to avert disaster, the work of building a national Green Growth Alliance to birth a Green New Deal won't be in vain. The effort to reinvent the system will build up important knowledge and establish invaluable relationships. Even in the most dire, "hard-landing" scenario, we can redeploy that wisdom and those networks to make the best of even the worst situations.

And as for me, I do not believe or accept that the fight for the future has already been lost. We should not let the possibility of eco-apocalypse paralyze us; we should let it motivate and propel us. Whatever our fate, we know one thing: hiding out or holding back won't save us.

Many of the best people in the country and the world have not been heard from yet. Many of the best ideas have not yet surfaced or been taken to scale. As the green wave encompasses more people, it will produce more innovation, inventiveness, and passion than we can possible appreciate or imagine. Just as there are unimaginable bad things on the horizon, there are also unimaginable good things. And I am betting on them.

And let's not forget that most people walking around in a mall or on a college campus are carrying on them better technology than the entire U.S. government had when it put a man on the moon. Each one of us is a walking technological superpower. We can access information and communicate instantly with people around the world. Given the capacities available to us, our wildest dreams and biggest hopes are probably too small.

In the coming crises, the only viable response will be collective action, supported by effective government. That approach will demand a different concept of leadership, even among people who are already focused on saving the world.

My friend Paul Hawken, an eco-entrepreneur, poet, and visionary, writes in his book *Blessed Unrest:* "Healing the wounds of the earth and its people does not require saintliness or a political party, only gumption and persistence. It is not a liberal or conservative activity; it is a sacred act." We need leaders with a comprehensive vision and positive agenda, who pursue the most cutting-edge approaches to education reform, health care, violence prevention, job creation, and eco-friendly economics.

Now is not the time to shrink from the challenge of saving our only home in the universe. Now is not the time to pull into ourselves, retreating into either a survivalist or an escapist mode. To the contrary, this is the time for titans, not turtles. Now is the time to open our arms, expand our horizons, and dream big. Big problems require big solutions. World-historic problems require world-class leadership. To prevail, we will need tens of thousands of heroes at every level of human society.

No one great leader can fix this nest of problems. Al Gore has been the planetary Paul Revere on this issue; he could have sat on his hands after 2000, growing increasingly bitter. Instead, he has used his fame to awaken the world to the threat of a climate catastrophe. If we survive, the entire human family will owe a great deal to him. But no one person can do it alone. We need thousands of Al Gores, thousands of Susan B. Anthonys, Rosa Parkses, Martin Luther Kings, César Chávezes, Franklin Roosevelts, and Eleanor Roosevelts. As always, those stalwart heroes will emerge from the ranks of ordinary people just like me and you. So each and every one of us should stop playing small and license ourselves to become one of the giants of the new century. We will need champions by the truckload.

If we stand for change, we can spark a popular movement with power, influence, magic, and genius. We won't just have the movement we have always wanted. We will have the country we have always wanted—and the world for which our hearts have longed.

Now is the time for us to raise our sights. Now is the time for America to dream again. Even in the midst of new dangers, now is the time for us to unshackle our imaginations. Let us envision meeting our economic and ecological challenges with our heads held high—not buried in our hands.

We can put "solution centers" in every town and neighborhood to train young workers in new technologies and ancient wisdom. We can envision our rural and urban youth creating zero-pollution products to sell. We can imagine formerly incarcerated people moving from jail cells to solar cells—helping to harvest the sun, heal the land, and repair their own souls. We can help local communities join hands—across lines of class and color—to honor the Earth, create new jobs, and reduce community violence.

We can create clear skies over our major port cities. Where idling ships once fouled the air, we can build solar-powered energy stations that let docking sea vessels power up cleanly. And we can send trucks using hybrid engines or cleaner biodiesel blends to take fair-trade goods off those ships without polluting the neighborhood.

We can build eco-industrial parks on land once blighted by prisons. We can pass laws that help transform our dying blue-collar towns, struggling rural regions, and poor neighborhoods into dignified, green-collar meccas. We can help our Rust Belt cities blossom as Silicon Valleys of green capital.

And we need not limit the innovation and industry of America to communities within our own borders. We can cooperate globally to give Africa, Latin America, and other developing regions the means to grow economically, while preserving their natural environments. U.S. entrepreneurs can help vast areas of China and India power up with clean energy. And we can learn from the world.

The best answer to our ecological crisis also responds to our socioeconomic crisis. The surest path to safe streets and peaceful communities is not more police and prisons, but ecologically sound economic development. And that same path can lead us to a new, green economy—one with the power to lift people out of poverty while respecting and repairing the environment.

We can lift the sword of war from over the heads of our sisters and brothers across the globe. We will set this country on the path of partnership with, not domination over, the world community. We can retrieve the Bill of Rights from the garbage bin. We can use our powers to heal the Earth, not pave it. We can deepen our nation's commitment to human rights for people of all races, religions, genders, and birthplaces.

These are our sacred duties. And we will meet them. When we do, the United States will, once again, be the leader of the whole world. But this time, not in war. Not in pollution. Not in incarceration rates.

Instead, we will lead the world in human rights and in social justice. In world-saving technologies and sustainable job creation. We will lead by showing the world how a strong, multiracial nation can unite itself to solve its toughest problems. That's where our new movement has the potential to take us all.

Some will call this unrealistic. They will advise America to keep her dreams small. But that cynicism is the problem, not the solution. A national commitment to green-collar jobs will renew this nation. We can take the unfinished business of America on questions of inclusion and equal opportunity and combine it with the new business of building a green economy—and thereby heal the country on two fronts and redeem the soul of the nation. The odds against achieving this kind of miraculous turnaround may seem daunting, but our forebears faced long odds—and overcame them. We can, too.

So I will close with the words and the example of the British prime minister Winston Churchill. As Adolf Hitler's monstrous shadow fell over the capitals of Europe, the prospects for democracy and human civilization had never looked bleaker. The United States was still refusing to fight. All other allies had fallen. And so Britain stood alone against the terror. Most observers doubted the British could last one week against Hitler's murderous onslaught.

But Churchill dared to position his lonely and isolated island nation as freedom's last barricade. On his first day in office, Churchill assigned himself the mission of blocking the advance of the fearsome Nazi death machine.³ He announced to the Parliament and the world: "I take up my task in buoyancy and hope. I feel sure that our cause will not be suffered to fail among men. I feel entitled at this juncture, at this time, to claim the aid of all and to say, 'Come then, let us go forward together with our united strength.'"⁴ His defiance and determination helped turn the tide of world affairs. And so it was that, against all odds and after great and horrendous suffering, the champions of democracy prevailed.

That generation—the so-called greatest generation—looked out and saw two futures: one with a blood-soaked Hitler ruling the world and another with the fascist threat forever eliminated. They decided to go out and do whatever it took to win that better future for the coming generations.

The following generation—the generation of the civil rights struggle and the women's liberation movement—looked out and saw two futures: one with Jim Crow and misogyny forever dividing our society and another with segregation torn from the pages of our law books and bigotry erased from the hearts of our children. They decided to go out and do whatever it took to win that better future for the coming generations.

Today, we are living in the world won for us by the sacrifices of those generations. It is far from perfect, but it is infinitely better than what might have been.

As a result of their heroism, we live in freer societies, with the liberty to act—or not to act. And now we stand at our own crossroads, looking out upon two futures: one with rising temperatures, rising oceans, and rising violence on a hot and strip-mined planet and another with expanding organic harvests, growing solar arrays, and deepening global partnerships on a green and thriving Earth. Given those same choices, we know in our hearts what our parents' generation—and our grandparents' generation—would have done for us. Come, then, let us go forward together with our united strength—and win that better future for the generations to come.

From Hope to Change: Governing Green

W HEN I FIRST proposed this book to HarperOne in the late summer of 2007, very few people had heard the term "green-collar job" or "green job." By the time the first edition went to press in summer 2008, the idea had become a mainstay in U.S. politics.

Rarely has an idea exploded into political consciousness so rapidly. At the end of 2007, Speaker Nancy Pelosi, Senator Bernie Sanders, Senator Hillary Clinton, U.S. Representative Hilda Solis, and U.S. Representative Mike Tierney worked together to pass the first ever Green Jobs Act. President George W. Bush signed it into law as a part of the 2007 energy package. In 2008, all three Democratic Party frontrunners for their party's presidential nomination embraced the concept. And in November of that year, U.S. Senator Barack Hussein Obama was elected president of the United States—with clean-energy jobs as the cornerstone of his platform.

Along the way, this little book became a surprise bestseller—and something of a manifesto for those seeking a fresh approach to our environmental and economic woes. As we prepare the book for a 2009 paperback release, I am about to join President Barack Obama's administration—as a special adviser for green jobs. (I will be the first person with "green jobs" in my title ever to serve in the White House.)

A PERFECT STORM PROPELS "GREEN JOBS" TO CENTER STAGE

The vertical trajectory of the concept is simply stunning, with few parallels in modern U.S. political history. Even doubters and detractors must admit that this novel idea—one that was championed by only a handful of true believers twenty-four months ago—is helping to reshape U.S. politics. More important, it is now poised to play a central role in the reinvention of the U.S. economy.

A number of factors converged to propel the idea onto center stage. The primary ones made jobs and energy prices two of the central issues of the 2008 presidential election. That development represented a tremendous change from 2004. Back then, the newly formed Apollo Alliance was promoting clean-energy jobs, but the topic was mostly a side issue. Democratic Party nominee John Kerry rarely mentioned it. By the 2008 election cycle, several factors had changed the equation: Hurricane Katrina, Al Gore's advocacy for climate solutions, the pain of the economic recession, and—during the summer—rising energy prices. Together, these circumstances moved the interrelated challenges of energy, environment, and the economy to ground zero in U.S. politics.

Even Republican Party nominee John McCain routinely celebrated the jobs that can be created in the clean-energy field. He is not alone in hoisting a green flag over the Grand Old Party. Conservative Texas oilman T. Boone Pickens has planted his own green stake in the ground—hundreds of them, in fact. He plans to install acres of wind turbines in an effort to promote clean energy as the best path to U.S. energy independence. Christian evangelicals—especially the younger ones—have embraced the idea of "creation care"; they see it as a faith-friendly on-ramp to environmental stewardship and activism. Archconservative Newt Gingrich has written a book called *A Contract with the Earth*.

The concept of "green jobs" is today embraced on TV talk shows, in think tanks, and in the halls of power. Today, legislation using the term is routinely introduced in school-board rooms, city halls, and state governing sessions across America. And yet in the brave new world of ecological solutions, the battle to move from rhetoric to reality has really just begun.

THE NEW FIGHT: "CLEAN GREENS" VERSUS "DIRTY GREENS"

The shifting landscape has created opportunities as well for the enemies of positive change. In the summer of 2008, with gas prices climbing above $4 a gallon, champions of the dirty-energy status quo launched a brilliant offensive to divert and derail the green movement. They advanced three clever slogans: "All of the Above"; "Drill Here, Drill Now, Pay Less"; and "Stop the War on the Poor."

The first is perhaps the most pernicious. The main defenders of the planet-cooking status quo have largely given up spreading confusion about whether global warming is a real problem. They lost that fight, at least among polite company. So their new tactic is to spread confusion not about the problem—but about the solutions. Today they are deliberately blurring distinctions between themselves and the champions of genuine answers to the problem of climate change.

In fact, under the tempting slogan "All of the Above," they have affirmatively adopted the call for clean-energy technology. And yet they have not abandoned any of their dirty-energy technologies. Instead, they now argue that America should do whatever it takes to lower energy prices, deploying the safest, cleanest energy solutions *and* the dirtiest, most dangerous energy solutions—all at the same time. This seemingly innocent slogan lets polluters stuff proposals for oil-shale development, tar-sand exploitation, and nuclear power into energy plans that also feature solar panels and wind turbines.

In other words, rather than fight the green wave, the proponents of eco-apocalypse have chosen to join it. They cannot succeed by flying their pirate flag of death. They refuse to show the white flag of surrender. So they have chosen simply to grab a green banner and hoist it alongside their true colors—and hope nobody notices.

In some ways, politicians are taking a page from "green-washing" corporations—the ones that change their ads, but neglect to change their wasteful processes or toxic products. Today we are seeing "green-washing" politicians who put solar panels in their speeches and wind turbines in their commercials, but keep their pro-polluter agendas. They say the right words, but their environmental commitment is no deeper than the green sheen on an oil slick. Call it the rise of the "dirty greens."

At least the second slogan is up-front and direct. GOP strategic mastermind Newt Gingrich has led the charge to open America's coastlines to offshore drilling with the mantra "Drill Here, Drill Now, Pay Less." He has skillfully used rising fuel prices to stoke public support for climate-destroying measures. Despite the reality that offshore drilling would take a decade to produce even a two-cent drop in gas prices, polls show that his slogan has been very effective.

The third slogan is perhaps the most tragic: "Stop the War on the Poor." Using this motto, a black-led, polluter-supported "coalition" is accusing environmentalists of being "punishers of the poor" for not allowing the drilling that the group alleges would reduce gas prices. The group's debut rally featured African Americans holding signs saying, "Environmental Groups Don't Feed My Family" and "Food or Fuel? Don't Make Me Choose." The backlash alliance of which I warned in these pages is no longer theory. It has arrived—full force.

So the polluters have proven savvy enough to nod in a green direction, while actually accelerating their climate-destroying agenda. Ironically, those of us who seek a green future have already won the argument for change. But we have not yet won enough public support, at this stage, to implement the kind of changes we need.

Fortunately, we still have time. The next rounds will be fought over the three P's: price, people, and the planet. The debate will be: What's the best strategy to reduce energy prices? What's the best economic strategy to help people? And in dead last place, but still more important than ever before: What's right ecologically for the planet, especially for the Earth's climate? The clean greens must convince the public that a true clean-tech revolution gives the best answer for all three. To succeed, we must shift dramatically in the direction of social-uplift environmentalism and eco-populism.

Cash-strapped, economically fearful families are emerging as the swing constituency on climate policy. The only way to draw them into the coalition for real solutions is by delivering fully on the promise of a green economy that provides increased work, more wealth, and better health for them and their children. Delivering on that promise is the great work of the new century.

OBAMA: FOCUS ON THE COLOR OF HIS DREAMS

Luckily, we have a leader in the White House who is committed to carrying out a clean-energy revolution as a key part of restoring America's economy. He speaks powerfully and eloquently about a new energy economy, one capable of producing millions of good jobs for Americans of every color and class.

Although my ideas do not match perfectly with his (an impossibility in any event), the points of overlap between the administration's goals and my own ideas are compelling and exciting. I am honored that I was asked to serve in Washington, D.C. I am proud that I will be able to add my voice, my ideas, and my labors to the Obama-Biden administration.

It will be a historic administration—and not just for the obvious reasons. Of course, President Obama has already won a place in the history books, because of the racial barriers he has smashed. But shattering the one barring the way to our nation's highest office is just a small part of the history that this rare leader has the opportunity and the obligation to make. I pray that ultimately it won't be the color of his skin that determines his legacy, but the color of his dreams. After all, we finally have a U.S. president whose dreams for America are green. And that is exactly what the world needs now, because during his time in office the United States will make vital decisions that may well determine whether humanity survives into the twenty-second century.

At the same time, even the greatest president cannot make changes on this scale all by himself. Lasting change requires more than just a great leader in the White House. Real change also requires great innovators within communities—and great movements in the streets. As President Obama sets America on a course toward ecological and economic sanity, all of us have a responsibility to help. In fact, as powerful as Obama will be as a "president," he may prove to be even more powerful as a "precedent": an inspirational example showing us all what is possible when people dare to act on the audacity of their own hopes. Only time will tell.

What I do know is this. At the dawn of the next century, if we have not cooked the planet, if we have not torn our globe apart in wars over energy and water, if there are millions of wind turbines turning and solar panels gleaming, if our forests are flourishing, if our weather is mild, if our skies are blue, if our children are safe, it will *not* be because Barack Obama was the first black president. It will be because Barack Obama was our first *green* president—and because all of us did everything we could to help him succeed.

So let us begin.

March 2009

Action Items

APPLY PRESSURE HERE

Check greenforall.org for up-to-date information on currently pending legislation that needs your support. Your e-mails and phone calls to Congress help tremendously in the ongoing battles to acquire full funding for green-collar job programs.

HAS YOUR MAYOR SIGNED THE LOCAL GOVERNMENT GREEN JOBS PLEDGE?

The following pledge was created by Green For All in partnership with the Apollo Alliance, Center for American Progress, and ICLEI–Local Governments for Sustainability.

On June 24, 2008, at the U.S. Conference of Mayors (USCM) annual meeting in Miami, 1,139 mayors from around the country passed a resolution to support this pledge.

AS LOCAL GOVERNMENT LEADERS, WE COMMIT TO:
Focus on green-collar jobs as a central strategy for advancing environmental, economic, and climate protection goals.

Green-collar jobs:

- Provide pathways to prosperity for all workers;
- Offer competitive salaries and lead to a lasting career track, thereby strengthening the U.S. middle class;
- Emphasize community-based investments that cannot be outsourced; and
- Contribute directly to preserving or enhancing environmental quality.

Grow an inclusive sustainable economy that creates green-collar jobs that:

- Strengthen and make further progress on our stated commitment to improving the environment in ways that grow both the green economy and green-collar jobs locally;
- Build on climate and environmental commitments to create market demand for green products, services, and skilled workers and create more prosperous local economies;
- Catalyze green-collar job creation and training by supporting policies that drive public and private investment in an inclusive local green economy; and
- Develop education and job-training programs that improve social equity and provide pathways out of poverty for our residents while strengthening our middle class by equipping workers for high-demand jobs in the green economy.

Execute tangible actions that place a priority on building an inclusive green economy that will:

- Involve our communities in developing and enacting green-collar jobs initiatives;

- Drive accountability and resolve to continuously improve and strengthen our efforts to invest in climate solutions that create economic opportunity and build sustainable communities;

- Provide accessible leadership that is responsive to our communities as we evolve the green economy;

- Use the purchasing power of our local governments to create markets for renewable energy, energy efficiency, and other green industries; and

- Invest new local government resources in programs and initiatives that build an inclusive green economy, while leveraging and aligning existing public resources, and private sources of capital and finance, toward these same goals.

We commit to join together as a movement of local governments across the United States to seize the economic, environmental, and social opportunities offered by building an inclusive green economy of high-quality jobs and a thriving green-collar workforce.

The pledge form can be downloaded at:
www.greenforall.org/files/Green%20Jobs%20Pledge%20Packet .pdf.

Resource List

NATIONAL GROUPS ADVOCATING FOR GREEN-COLLAR JOBS

1Sky
Community dedicated to aggregating a massive nationwide movement by communicating a positive vision and a coherent set of national policies that rise to the scale of the climate challenge.
www.1sky.org

American Council on Renewable Energy (ACORE)
Council working to bring all forms of renewable energy into the mainstream of America's economy and lifestyle.
www.acore.org

Apollo Alliance
Coalition of business, labor, environmental, and community leaders working to catalyze a clean-energy revolution in America to reduce our nation's dependence on foreign oil, cut the carbon emissions that are destabilizing our climate, and expand opportunities for American businesses and workers.
www.apolloalliance.org

Applied Research Center
Center advancing racial justice through research, advocacy, and journalism.
www.arc.org

Blue Green Alliance
Partnership between the United Steelworkers and Sierra Club to spur dialogue between environmentalists and labor on topics of global warming, clean energy, toxics, and fair trade.
www.bluegreenalliance.org

Center for American Progress
Progressive think tank dedicated to improving the lives of Americans through ideas and action.
www.americanprogress.org

Center for State Innovation
Center that helps governors and other state executives advance and implement innovative, progressive policies that better the lives of the people they serve.
www.stateinnovation.org

Center on Wisconsin Strategy (COWS)
National policy center and field laboratory for high-road economic development—a competitive market economy of shared prosperity, environmental sustainability, and capable democratic government.
www.cows.org

Color of Change
Organization that exists to strengthen black America's political voice. Its goal is to empower members—black Americans and their allies—to make government more responsive to the concerns of black Americans and to bring about positive political and social change for everyone.
www.colorofchange.org

Community Food Security Coalition
Coalition dedicated to building strong, sustainable local and regional food systems that ensure access to affordable, nutritious, and culturally appropriate food by all. It seeks to develop self-reliance among all communities

in obtaining their food and to create a system of growing, manufacturing, processing, making available, and selling food that is regionally based and grounded in the principles of justice, democracy, and sustainability.
www.foodsecurity.org

Energy Action Coalition
Coalition of more than forty organizations from across the United States and Canada, founded and led by youth, to support and strengthen the student and youth clean-energy movement in North America. Its Campus Climate Challenge leverages the power of young people to organize on college campuses and high schools to win 100 percent clean-energy policies at their schools.
www.energyactioncoalition.org

The Engage Network
Network that trains leaders to create self-replicating small groups that both take care of people and change the world at the same time. It is creating small circles, including discussion groups and curricula based upon this book in partnership with Green For All.
www.engagenet.org

Green For All
National advocacy organization working to build an inclusive green economy strong enough to lift people out of poverty. It shapes debate and shares best practices, spurs action in the federal government and the private sector to ensure that the United States has an abundant supply of well-trained "green-collar" workers and entrepreneurs, focusing on those from disadvantaged backgrounds.
www.greenforall.org

Intertribal Council on Utility Policy (Intertribal COUP)
Council involved in policy issues and outreach education to Tribal governments, Tribal Colleges, and indigenous environmental organizations on telecommunications, climate change, energy planning, energy efficiency, and renewable energy development. The policy work includes specific proposals to support renewable energy development and energy efficiency.
www.intertribalcoup.org

Lifestyles of Health and Sustainability (LOHAS)
Group of companies practicing "responsible capitalism" by providing goods and services using economic and environmentally sustainable business practices. An annual conference covers industry trends and how to run a successful LOHAS business.
www.lohas.com/

Local Governments for Sustainability
Association of local governments that provides tools and technical assistance to local governments to set and achieve their climate protection and sustainability goals.
www.iclei-usa.org

Reconnecting America and the Center for Transit-Oriented Development
National nonprofit organization that is working to integrate transportation systems and the communities they serve, with the goal of generating lasting and equitable public and private returns, providing people with more housing and mobility choices, improving economic and environmental efficiency, providing concrete, measurable solutions to reducing greenhouse-gas emissions and reducing dependence on foreign oil.
www.reconnectingamerica.org

Transportation Equity Network
National coalition reforming unjust and unwise transportation and land use policies. It works to win funding for transportation and job-training programs.
www.transportationequity.org

U.S. Green Building Council
Nonprofit dedicated to sustainable building design and construction. It established national green building standards, the LEED standards.
www.usgbc.org

The Workforce Alliance
National coalition of community-based organizations, community colleges, unions, business leaders, and local officials advocating for public policies that invest in the skills of America's workers, so they can better

support their families and help American businesses better compete in today's economy.
www.workforcealliance.org

LOCAL, REGIONAL AND NATIONAL GROUPS ORGANIZING GREEN-COLLAR JOBS TRAINING OPPORTUNITIES

African American Environmentalist Association
National nonprofit that works to clean up neighborhoods by implementing toxics education and energy, water, and clean-air programs. It includes an African American point of view in environmental policy decision making and resolves environmental racism and injustice issues through the application of practical environmental solutions.
www.aaenvironment.com

Alameda County Career Center, East Oakland
California county agency providing free job training in the construction trades and other areas to those who qualify.
www.eastbayworks.com/ebw-resources/Oaklandeast.htm

Appalachian Voices
Regional nonprofit that works with communities across Appalachia to restore the region's forests and tackle the dual problems of mountaintop-removal coal mining and the construction of new coal-fired power plants.
www.appvoices.org

Asian Neighborhood Design (AND)
San Francisco nonprofit offering a free preapprenticeship construction training program. Students are automatically indentured in the carpenters union upon completion of the course and provided with a small stipend and tool kit.
www.andnet.org

Bay Area Construction Sector Intervention Collaborative
Regional nonprofit collaborative of service providers and job-training agencies promoting economic self-sufficiency by increasing access to career-path jobs in the construction industry.
www.turner-oak.com/workforce.cfm

B'more Green
Nonprofit providing disadvantaged Baltimore residents with environmental job training and employment development initiatives in the
areas of brownfield remediation and hazardous-materials abatement and
containment.
www.civicworks.com/bmghome.html

Border Ecology Project
Regional nonprofit that develops solutions to environmental and health
problems in the U.S.-Mexico border and other Latin American regions
through collaboration with nongovernmental organizations, academic
institutions, private consultants, and government agencies. It adds
community input to national and international policy discussions and
negotiations.
www.borderecoweb.sdsu.edu/bew/drct_pgs/b/bep.html

Boston Connects, Inc.
Boston nonprofit implementing an array of community and economic
development initiatives designed to improve communities in Boston that
have experienced chronic divestment over the years.
www.cityofboston.gov/bra/bostonez/index.html

Build It Green
California statewide nonprofit that provides a two-day (sixteen-hour)
Certified Green Building Professional Training covering all aspects of
green building: energy, water, materials, indoor air quality, and implementing green in a company's operations and marketing.
www.builditgreen.org

Building Opportunities for Self-Sufficiency
Alameda County, California, nonprofit offering free job training for the
homeless and formerly homeless focusing on building maintenance, computer, and culinary skills.
www.self-sufficiency.org

**Business Alliance for Local Living Economies
(BALLE)**
National network that creates, strengthens, and connects local business networks dedicated to building strong local living economies.
www.livingeconomies.org

Center for Environmental Policy and Management at the University of Louisville
University center focusing on three project areas: (1) brownfields/smart-growth research, (2) environmental policy and forecasting, and (3) the EPA Region 4 Southeast Regional Environmental Finance Center.
www.cepm.louisville.edu

Center for Integrated Waste Management, Department of Civil, Structural and Environmental Engineering, State University of New York at Buffalo
University center that promotes the development and application of improved technologies and management methods for (1) more effectively remediating past environmental contamination and promoting redevelopment of formerly contaminated properties, and (2) preventing, reducing, reusing, and recycling industrial and municipal waste streams.
www.ciwm.buffalo.edu

Communities for a Better Environment
California statewide nonprofit doing community organizing in working-class communities of color who are bombarded by pollution from freeways, power plants, oil refineries, seaports, airports, and chemical manufacturers.
www.cbecal.org

DC Greenworks
Washington, D.C., business that provides full-service green-roof design, installation, and consulting. All of its contracts serve to train and employ underserved adults in the skills necessary to meet the growing demand for these new environmental services and technologies.
www.dcgreenworks.org

East Bay Conservation Corps
Nonprofit providing free job training for East Bay youth based on environmental stewardship and community service.
www.ebcc-school.org

Ella Baker Center for Human Rights
A nonprofit strategy and action center, based in Oakland, California, working for justice, opportunity, and peace in urban America. It promotes positive alternatives to violence and incarceration through four cutting-edge campaigns.
www.ellabakercenter.org

Federation of Southern Co-ops, Land Assistance Fund
Southern regional nonprofit operating a rural training and research center. It assists in land retention and development.
www.federationsoutherncoop.com

Food from the Hood
Los Angeles nonprofit fostering business, academic, and life skills for challenged young people. It provides tutoring, college entrance-exam training, mentoring, and business-skills development; it is committed to supporting students in business ventures that are socially responsible, environmentally sound, and neighborhood-friendly.
www.foodfromthehood.com

The Garden Project
San Francisco nonprofit that supports former offenders by providing training in horticulture skills. Participants grow organic vegetables that feed seniors and families in San Francisco.
www.gardenproject.org

Greater Philadelphia Urban Affairs Coalition
Philadelphia nonprofit that unites government, business, neighborhoods, and individual initiatives to improve the quality of life in the region, build wealth in urban communities, and solve emerging issues. It offers extensive workforce training.
www.gpuac.org

Green Communities Online
National nonprofit that has a five-year, $555 million commitment by Enterprise to build 8,500 healthy, efficient homes for low-income people and make environmentally sustainable development the mainstream in the affordable housing industry.
www.greencommunitiesonline.org

Green Worker Cooperatives
Bronx, New York, nonprofit that incubates worker-owned and environmentally friendly cooperatives in the South Bronx.
www.greenworker.coop

Greencorps Chicago
City-administered horticultural and green-industries paid job-training program with wraparound and job placement services.
www.cityofchicago.org/Environment

Grid Alternatives
San Francisco Bay Area nonprofit that trains community volunteers in the theory and practice of solar electric installation and provides them with hands-on experience with solar-installation projects.
www.gridalternatives.org

Groundwork Providence
Providence, Rhode Island, nonprofit that provides educational environmental programs such as the Providence Neighborhood Planting Program, Spring and Fall Clean-Ups, Environmental Education Clubs, Brownfields Job Training Program, and Summer Green Teams.
www.groundworkprovidence.org

Groundwork USA
East Coast regional nonprofit network of independent, not-for-profit environmental businesses that work with communities to improve their environment, economy, and quality of life through practical local projects.
www.groundworkusa.net

Growing Home

Chicago nonprofit that provides job training and employment opportunities for homeless and low-income people within the context of an organic agriculture business.
www.growinghomeinc.org

Houston Community College

Texas community college offering associate degrees, certificates, workforce training, and lifelong learning opportunities.
www.hccs.edu

Impact Services Corporation

Philadelphia corporation enabling people in need to attain the hope, motivation, and skills necessary to reach their fullest potential and achieve personal and family self-sufficiency.
www.pacdc.org/cgi-bin/board.cgi?ISC

Isles: Fostering Self-Reliance

Trenton, New Jersey, nonprofit operating activities that recognize the interdependence of physical, economic, health, and social development strategies to address the problems of distressed communities, with the mission of fostering more self-reliant families in healthy, sustainable communities.
www.isles.org

LA City College, CalWORKs program

Community college that partners with the Department of Public Social Services (DPSS) to provide education and successful transition from welfare to work.
www.lacitycollege.edu

Lao Family Community Development, Inc.

California nonprofit that provides free job training in construction trades, health care, building maintenance, and other areas, with a focus on South East Asian refugee and immigrant communities.
www.laofamilynet.org

Los Angeles Conservation Corps
LA nonprofit that provides at-risk young adults with job training, education, and work-skills training with an emphasis on environmental and service projects that benefit the community. It offers a "clean and green" program.
www.lacorps.org

Miami-Dade Community College
Community college providing accessible, affordable, high-quality education, in partnership with the dynamic, multicultural community.
www.mdc.edu

Milwaukee Community Service Corps
Milwaukee nonprofit that provides participants with the opportunity to learn new skills, earn a wage, serve their community, earn a high-school equivalency diploma, and prepare themselves for post-corps college or trade apprenticeships. In the field, corps members renovate vacant homes, plant community gardens, landscape vacant lots, remove graffiti, intern in youth-service agencies, perform lead outreach and reduction activities, distribute food for food pantries, engage in recycling projects, and construct new playgrounds.
www.milwaukeecommunityservicecorps.org

Mo' Better Food
Oakland, California, nonprofit promoting activities that bring generations together and encourage healthy economic development by increasing community leadership, job training, entrepreneurship, community pride, and community ownership.
www.mobetterfood.com

Montana Tech of the University of Montana
University program selected by the EPA for a Brownfields Job Training and Development Demonstration Pilot, in partnership with Crow Nation. The job training pilot will focus on residents of the Crow Indian Reservation (7,900 tribal members).
www.mtech.edu

Mothers on the Move
Bronx, New York, nonprofit that operates community campaigns for decent housing, traffic safety, and environmental justice, including renovated buildings, redeveloped and new parks, and safer streets.
www.mothersonthemove.org

Native Movement
Southwest regional nonprofit that supports projects and campaigns led by youth that are focused on peace, sustainability, youth leadership development, healing, community building, and movement building.
www.nativemovement.org/southwest/

New Jersey Youth Corps
Government agency providing full-time instructional and community service programs for school dropouts, with the completion of a high-school curriculum and employment as the ultimate goals for each student. Students receive academic instruction and perform community service work. A one-month orientation that includes academic and interest/aptitude assessment is followed by placement in community service work crew projects and continuation in basic skills classes.
www.ed.gov/pubs/EPTW/eptw14/eptw14c.html

New York City Environmental Justice Alliance
Nonprofit that supports the work of member groups based in low-income communities throughout New York City. It coordinates advocacy efforts, facilitates networking, and helps to replicate successful projects and activities.
www.nyceja.org

Oakland Private Industry Council
Oakland, California, nonprofit that helps the community maintain no-fee career centers and workforce-development programs. Its goal is to aid the economy by helping job seekers prepare for work, then providing employers with highly trained employees.
www.oaklandpic.org

Office of Applied Innovations
Chicago nonprofit that enhances the capacity of underserved individuals and their communities to contribute significantly to social-environmental

equity, equal access to educational and employment opportunities, and economic self-sufficiency and self-determination.
www.oaiinc.org

Oregon Tradeswomen, Inc.
Regional nonprofit that promotes success for women in the trades through education, leadership, and mentorship.
www.tradeswomen.net

Pacific Energy Center
Center, run by Pacific Gas and Electric Company Corporation, that provides education in energy conservation and green building techniques, systems, and materials.
www.pge.com/pec/

People's Grocery
Oakland, California, nonprofit offering free job training in building a local food system and improving the health and economy of the community.
www.peoplesgrocery.org

Pivotal Point Youth Services
Oakland, California, nonprofit that provides free intensive employment training, vocational skills development, entrepreneurship training, case management, and other comprehensive supportive services for youth ages sixteen to twenty-four.
www.ppys.org

Regional Technical Training Center
San Francisco Bay Area nonprofit providing practical training, resumé and interviewing skills assistance, and job placement.
www.rttc.us

Rising Sun Energy Center
Berkeley, California, nonprofit that provides a demonstration site and education center for renewable energy and conservation techniques.
www.risingsunenergy.org

Riverside New Visions Program
Community college providing eligible welfare recipients the opportunity to attend Riverside Community College in order to receive training in basic skills (computer literacy, English, and math) and enroll in an occupational mini-program. The goal is to help students climb the "career ladder" to higher-paying and more satisfying jobs.
www.rcc.edu

Second Chance
Non-profit working with local and regional architects, builders, and contractors to find old buildings entering the demolition phase and rescue all reusable elements in the Baltimore, Philadelphia, and Washington, D.C., areas. It offers low-income residents training in a wide variety of skill sets ranging from carpentry to craftsmanship.
www.secondchanceinc.org

Solar Richmond
Richmond, California, nonprofit promoting the use of solar power and energy efficiency in order to bring the economic benefits of the green economy to the community. It serves the underemployed by educating a green-collar workforce and opening doors to employment.
www.solarrichmond.org

Southwest Network for Environmental and Economic Justice
Regional nonprofit that strengthens the work of local organizations and empowers communities and workers to affect policy on environmental and economic justice issues as these impact people of color in the southwestern United States and along the border region of Mexico.
www.sneej.org

St. Louis Community College
Community college focusing on residents of economically depressed communities in St. Louis, Missouri, and East St. Louis, Illinois. It furthers brownfields-related redevelopment activities spurred by EPA assessment and revolving loan fund pilots along with recently initiated efforts by the city to address lead and asbestos problems.
www.stlcc.edu

St. Nicholas Neighborhood Preservation Corps.
Brooklyn, New York, nonprofit that spearheads revitalization and sustainability of the multiethnic community.
www.stnicksnpc.org

STRIVE: Boston Employment Service
Boston nonprofit that removes barriers to employment including lack of money, offering programs at no cost to clients. Its core program trains participants within four weeks and gets them into paid employment quickly, with two years of follow-up support.
www.bostonstrive.org

STRIVE: East Harlem Employment Service
Harlem, New York, nonprofit that removes barriers to employment and offers free job-readiness training.
www.strivenewyork.org

Sustainable Economic Enterprises of Los Angeles
Nonprofit that promotes self-sustaining community and economic development activities within the city of Los Angeles, including sustainable food systems, social and cultural programs, and economic revitalization projects.
www.see-la.org

Sustainable South Bronx
Nonprofit that addresses land-use, energy, transportation, water, and waste policy and education in order to advance the environmental and economic rebirth of the South Bronx and to inspire solutions in areas like it across the nation and around the world.
www.ssbx.org

Texas Engineering Extension Service
University service that provides, through contracts and agreements with governments and overseas companies, unique specialized training and technical assistance to workers worldwide, ranging from the smallest volunteer fire departments to some of the largest companies in the world.
www.teex.com

Tradeswomen Inc. (TWI)
National nonprofit that targets outreach, supportive services, mentoring and networking for women interested in the construction trades.
www.tradeswomen.org

Turtle Mountain Community College
An autonomous Indian-controlled community college on the Turtle Mountain Chippewa Reservation focusing on general studies, undergraduate education, vocational education, direct scholarly research, and continuous improvement of student learning.
www.turtle-mountain.cc.nd.us/

Urban Habitat
Alameda County, California, regional nonprofit that builds bridges between environmentalists, social justice advocates, government leaders, and the business community. Its work has helped to broaden and frame the agenda on toxic pollution, transportation, tax and fiscal reform, brownfields, and the nexus between inner-city disinvestments and urban sprawl.
www.urbanhabitat.org

West Virginia University Research Corporation
Nonprofit providing evaluation, development, patenting, management, and marketing services for inventions of the faculty, staff, and students of the university. The corporation also serves as the fiscal agent for sponsored programs on behalf of the university.
www.research.wvu.edu/wvu_research_corporation

The Workplace, Inc.
Southwestern Connecticut regional nonprofit providing a seamless, coordinated system of education, training, and employment that is easily accessible and that meets the needs both of employers and of persons who face barriers to good employment. It collaborates with business, education, government, and community agencies that include economic development, employment and training, and human services.
www.workplace.org

Young Community Developers, Inc.
San Francisco nonprofit that prepares Bayview–Hunters Point and San Francisco residents, particularly those with barriers to employment, for current and future workforce needs.
www.ycdjobs.org

The Youth Employment Partnership, Inc.
Nonprofit providing employment training to Oakland, California, youth.
www.yep.org

RENEWABLE ENERGY TECHNOLOGY, PRODUCTS, AND SERVICES

Austin Energy
Blue Sun Biodiesel
Bonneville Environmental
  Foundation
California Cars Initiative
Cape Wind Associates
Clean Air Now
Clean Edge
Clean Energy Group
Community Energy, Inc.
Community Fuels
Conergy, Inc.
Cooperative Community Energy
  Corporation
Daystar Technologies
East Haven Wind Farm
EcoLogical Solutions
Energy Innovations/Idea Labs
Environ Corporation
Environmental Energy Solutions
Global Resource Options, Inc.
Hawaii PV Coalition
HyGen Industries
McKenzie Bay International
Native Energy, LLC

NativeWind
Northern Power Systems
NRG Systems
Olympia Green Fuels
OurEnergy
Pacific Ethanol
Pioneer Valley Photovoltaics
PPM Energy
Pro Vision Technologies, Inc.
PV Powered
Renewable Energy Access
Sharp Solar
Shepherd Advisors
Solectria Renewables
Sterling Planet, Inc.
Sunlink, LLC
3 Phases Energy Services
United Bio Lube, Inc.
US Renewables Group
Wilson Turbo Power, Inc.

Energy Efficiency

Burlington Electric Department
Conservation Services Group

Engage Networks, Inc.
Johnson Controls
Kinsley Power Systems
Lightly Treading, Energy & Design
McKinstry
Vermont Energy Investment
  Corporation
Virent Energy Systems
Washington Electric Cooperative

Green Buildings, Infrastructure

All American Home Center
Blue Wave Strategies
Center for Smart Energy

CTO, NatureWorks (formerly
  Cargill-Dow)
Duce Construction Corporation
Ervin & Company
High Performance Building Tech-
  nology Team
Mazria Architects
Michigan Manufacturing Technol-
  ogy Center
NYC Transit
Quantec, LLC
Schultz Development Group, LLC
William Maclay Architects &
  Planners
William McDonough Partners

Notes

INTRODUCTION: REALITY CHECK
1. In July 2007, the National Petroleum Council released a report, entitled *Facing the Hard Truths About Energy,* that addressed the future scarcity of fossil fuels: "It's a hard truth that the global supply of oil and natural gas from the conventional sources relied upon historically is unlikely to meet projected 50 to 60 percent growth in demand over the next 25 years."
2. Russell Gold and Ann Davis, "Oil Officials See Limit Looming on Production," *Wall Street Journal,* November 19, 2007, http://online.wsj.com/article/SB119543677899797558.html.
3. *Crude Oil: The Supply Outlook,* an October 2007 report by the Energy Watch Group, conducts a meta-analysis of oil industry papers to arrive at the number of proven remaining reserves. The conclusion: peak oil has arrived. http://www.energywatchgroup.org/fileadmin/global/pdf/EWG_Oilreport_10-2007.pdf.
4. More information about the extraction of bituminous sands, better known as tar or oil sands, can be found on www.treehugger.com at http://www.treehugger.com/files/2006/01/alberta_tar_san.php. Oil-shale drilling in parts of Colorado, Utah, and Wyoming will drain scarce water resources, threaten habitats, and increase air pollution, according to a June 2007 report by the Natural Resources Defense Council (NRDC), *Driving It Home: Choosing the Right Path for Fueling North America's Transportation Future,* http://www.nrdc.org/energy/drivingithome.pdf.
5. Defenders of Wildlife estimates the economic impact of offshore drilling in Alaska's Bristol Bay, based on disruption of the local sustainable commercial fishing industry and related tourism, http://www.defenders.org/newsroom/press_releases_ folder/2008/04_08_2008_offshore_drilling_could_destroy_bristol_bay_fisheries.php.

228   *Notes*

6. Liquid coal results in double the carbon dioxide emissions: first during production, then again from the tailpipe, according to the NRDC report *Driving It Home,* http://www.nrdc.org/energy/drivingithome.pdf.

7. Intergovernmental Panel on Climate Change (IPCC), *Climate Change 2007: Synthesis Report,* November 2007, http://www.ipcc.ch/pdf/assessment-report/ar4/syr/ar4_syr_spm.pdf.

8. Speech by Susan Hockfield, President of MIT, at the American Association for the Advancement of Science (AAAS) annual conference, http://web.mit.edu/hockfield/speeches/2008-aaas.html.

9. American Solar Energy Society, "Renewable Energy and Energy Efficiency: Economic Drivers for the 21st Century," 2007, http://www.ases.org/ASES-JobsReport-Final.pdf.

10. Daniel B. Wood and Mark Clayton, "California Takes Lead in Global-Warming Fight," *Christian Science Monitor,* September 1, 2006, http://www.csmonitor.com/2006/0901/p01s01-usgn.html.

11. *Biofuels: At What Cost?* an October 2006 report by the Global Subsidies Initiative and the International Institute for Sustainable Development (IISD), reviews the many types of incentives in place to support the ethanol and biodiesel industry. http://www.globalsubsidies.org/files/aseets/pdf/Brochure_-_US_Report.pdf.

For example, a $71 million, 20-million-gallon-per-year ethanol plant under construction in Ohio lined up the following sources of public support: a $500,000 United States Department of Agriculture grant; $600,000 in Appalachian Regional Commission grants; $40,000 in training funds from the Ohio Department of Development; $400,000 in 629 Roadwork Development funds from ODOD; a $7,000,000 Ohio Water Development Authority loan; a $600,000 Rural Pioneer loan; and $36,261,024 in Ohio Air Quality Development Authority Revenue Bonds.

12. In 2006, over 14 percent of corn grown in the United States went into ethanol production. Jerry Taylor and Peter Van Doren, "Ethanol Makes Gasoline Costlier, Dirtier," *Chicago Sun-Times,* January 27, 2007, http://www.cato.org/pub_display.php?pub_id=7308.

13. Depending on demand, the world's supply of uranium will last only thirty to sixty more years, making nuclear a nonrenewable energy source, http://timeforchange.org/pros-and-cons-of-nuclear-power-and-sustainability.

14. The idea is to recapture the carbon dioxide produced by burning fossil fuels by photosynthesis: growing algae on the exhaust gas. "The result could then be turned into biodiesel (since many species of algae store their food reserves as oil), or even simply dried and fed back into the power station," according to a July 3, 2007, Economist.com article, "Old Clean Coal," http://www.economist.com/science/tq/displaystory.cfm?story_id=9431233.

15. David Adam, "Plan to Bury CO2," *Guardian,* Friday, September 5, 2003, http://www.guardian.co.uk/science/2003/sep/05/sciencenews.science.

16. R. Margolis, J. Zuboy, and the National Renewable Energy Laboratory, *Nontechnical Barriers to Solar Energy Use: Review of Recent Literature,* September 2006, http://www.nrel.gov/docs/fy07osti/40116.pdf.

17. David Bielo, "Combating Climate Change: Building Better, Wasting Less," *Scientific American,* May 18, 2007, http://www.sciam.com/article.cfm?id=combating-climate-change-building-better-wasting-less.

18. Thomas L. Friedman, *Hot, Flat, and Crowded: Why We Need a Green Revolution—and How It Can Renew America* (New York: Farrar, Straus and Giroux, 2008).

19. President Jimmy Carter said: "In little more than two decades we've gone from a position of energy independence to one in which almost half the oil we use comes from foreign countries, at prices that are going through the roof. Our excessive dependence on OPEC has already taken a tremendous toll on our economy and our people. This is the direct cause of the long lines, which have made millions of you spend aggravating hours waiting for gasoline. It's a cause of the increased inflation and unemployment that we now face. This intolerable dependence on foreign oil threatens our economic independence and the very security of our nation. . . . To give us energy security, I am asking for the most massive peacetime commitment of funds and resources in our nation's history to develop America's own alternative sources of fuel. . . . These efforts will cost money, a lot of money, and that is why Congress must enact the windfall profits tax without delay. It will be money well spent. Unlike the billions of dollars that we ship to foreign countries to pay for foreign oil, these funds will be paid by Americans, to Americans. These will go to fight, not to increase, inflation and unemployment." From the speech "Energy and the National Goals: A Crisis of Confidence," delivered July 15, 1979, http://www.americanrhetoric.com/speeches/jimmycartercrisisofconfidence.htm.

20. "Congress enacted the Crude Oil Windfall Profit Tax Act in 1980 in response to the excessive windfall profits that oil producers were earning following the deregulation of oil prices. The Act imposed a windfall profit tax on domestically produced crude oil that ranged from 30 to 70 percent of the producer's windfall profit. This act provided a tax credit of $3 per barrel of tar sands oil-equivalent to producers of alternative energy sources. Congress intended to use a substantial portion of the revenues from the windfall profit tax of crude oil to finance the tax credit for alternative fuels. The Omnibus Trade and Competitiveness Act of 1988 repealed the Crude Oil Windfall Profit Tax Act." Robyn Kenney, "Crude Oil Windfall Profit Tax Act of 1980, United States," *Encyclopedia of Earth,* http://www.eoearth.org/article/Crude_Oil_Windfall_Profit_Tax_Act_of_1980,_United_States.

CHAPTER 1: THE DUAL CRISIS

1. Larry Bradshaw and Lorrie Beth Slonsky, "Hurricane Katrina—Our Experiences," September 1, 2005, http://www.emsnetwork.org/cgi-bin/artman/exec/view.cgi?archive=56&num=18427. Although Bradshaw and Slonsky were not New Orleans residents, their story highlighted the compound system failures more comprehensively than any other first-person account I could find.

The fierce Mrs. Phyllis Montana Leblanc, featured in the documentary film *When the Levees Broke* by Spike Lee, composed a poem that better conveys the trauma of a N'awlins family divided by the disaster:

Not just the levees broke—
the spirit broke, my spirit broke
the families broke apart—

I want my momma back,
I want my sister back,
I want my nephew back.
The auction block broke
from so many African American bodies.
The sense of direction was broken
because of the darkness.
There was light from time to time but then?
They broke away and left us.
My being together broke
when I fell apart
The smell broke away from my skin
when I came out of the waters,
the waters that came and stood,
still, with the bodies of my people.
The dogs, shit, piss, rats, snakes, "unheard of" alligators
The broken smiles, the broken minds, broken lives.
And you know something?
Out of all of this brokenness
I have begun to mend
with God, and my deep, deep commitment
to the infinite strength, to never give up.
I am mending
I am coming back
God willing, for a long, long time.
So when you see the waters—
when you see the levees breaking
know what they really broke along with them.

2. Ross Gelbspan, "Hurricane Katrina's Real Name," *International Herald Tribune,* August 31, 2005, http://www.iht.com/articles/2005/08/30/opinion/edgelbspan .php.

3. Emmanuel Saez with Thomas Piketty, "Income Inequality in the United States, 1913–2002," Oxford University Press, 2007, http://www.elsa.berkeley.edu/~saez/ piketty-saezOUP04US.pdf.

4. Economic Policy Institute (EPI), *The State of Working America 2006/2007,* Table 5.1, http://www.stateofworkingamerica.org/tabfig/05/SWA06_Tab5.1.jpg.

5. United for a Fair Economy, *Executive Excess 2006,* http://www.faireconomy.org/ press/2006/ee06_ceos_pocket_the_spoils_preview.html.

6. Kevin Danaher et al., *Building the Green Economy* (Sausalito, CA: PoliPoint Press, 2007), p. 4.

7. U.S. Census Bureau, *Income, Poverty and Health Insurance Coverage in the United States: 2006,* http://www.census.gov/prod/2007pubs/p60-233.pdf.

8. Economic Policy Institute (EPI), Briefing Paper #171 "Trade Deficits and Manufacturing Job Loss," March 14, 2006, http://www.epi.org/content.cfm/bp171.

9. U.S. Census Bureau, *Health Insurance Estimates from the U.S. Census Bureau: Background for a New Historical Series,* June 2007, http://www.census.gov/hhes/www/hlthins/usernote/revhlth_paper.pdf.

10. Jeanne Sahadi, "President Signs Bankruptcy Bill," www.cnnmoney.com, April 20, 2005, http://money.cnn.com/2005/04/20/pf/bankruptcy_bill.

11. Jonathan Weil, Bloomberg.com, June 18, 2008, http://www.bloomberg.com/apps/news?pid=20601039&refer=columnist_weil&sid=a70JZmfcakF0. In the Bear Stearns bailout, the Federal Reserve is lending $29 billion to a Delaware limited-liability company that will hold a portfolio of illiquid Bear Stearns assets. JPMorgan Chase, which completed its purchase of Bear Stearns this month, will lend the Delaware entity $1 billion and absorb the first $1 billion of any losses. The Fed is on the hook for the rest.

12. U.S. Census Bureau, *Income, Poverty and Health Insurance Coverage in the United States.*

13. U.S. Census Bureau, *Income, Poverty and Health Insurance Coverage in the United States.*

14. Meizhu Lui et al., *The Color of Wealth* (New York: New Press, 2006), p. 34.

15. U.S. Census Bureau, *Educational Attainment in the United States: 2007,* http://www.census.gov/population/www/socdemo/education/cps2007.html.

16. U.S. Census Bureau, *Income, Poverty and Health Insurance Coverage in the United States.*

17. American Civil Liberties Union (ACLU), *Race & Ethnicity in America: Turning a Blind Eye to Injustice,* December 2007, http://www.aclu.org/pdfs/humanrights/cerd_full_report.pdf.

18. Federation for American Immigration Reform, "How Many Illegal Aliens?" http://www.fairus.org/site/PageServer?pagename=iic_immigrationissuecentersb8ca.

19. Applied Research Center and Northwest Federation of Community Organizations, Closing the Gap: Solutions to Race-Based Health Disparities, June 2005, www.thepraxisproject.org/tools/ClosingGap.pdf.

20. Life expectancy statistics from a presentation by Dean Robinson, University of Massachusetts Amherst, at a symposium entitled *Challenges of Black Economic Development: Discrimination and Access Issues,* March 23, 2005, http://www.5clir.org/SlaveryTranscripts/Session%204ABCD.htm.

21. Drug Policy Alliance Network, http://www.drugpolicy.org/communities/race/.

22. Pew Charitable Trust, *One in 100: Behind Bars in America 2008,* http://www.pewcenteronthestates.org/report_detail.aspx?id=35904.

23. A 2008 report by the National Priorities Project based on Defense Department data linked recruiting data to zip codes and median incomes and found that low- and middle-income families are supplying far more Army recruits than families with incomes greater than $60,000 a year. http://www.nationalpriorities.org/militaryrecruiting2007.

24. Intergovernmental Panel on Climate Change (IPCC), *Climate Change 2007: Synthesis Report,* November 2007, http://www.ipcc.ch/pdf/assessment-report/ar4/syr/ar4_syr_spm.pdf.

25. IPCC, *Climate Change 2007.*

26. IPCC, *Climate Change 2007.*

27. Alex Kirby, "Water Scarcity: A Looming Crisis?" BBC News Online, October 19, 2004, http://news.bbc.co.uk/1/hi/sci/tech/3747724.stm.

28. IPCC, *Climate Change 2007.*

29. IPCC, *Climate Change 2007.*

30. IPCC, *Climate Change 2007.*

31. Patricia Reaney, "Climate Change Raises Risk of Hunger," *Reuters,* September 5, 2005.

32. IPCC, *Climate Change 2007.*

33. IPCC, *Climate Change 2007.*

34. International Federation of Red Cross (IFRC), *World Disasters Report 1999* (Geneva: IFRC, 1999).

CHAPTER 2: THE FOURTH QUADRANT

1. M. Kat Anderson, *Tending the World: Native American Knowledge and the Management of California's Native Resources* (Berkeley and Los Angeles: University of California Press, 2005), p. 56.

2. Thom Hartmann, *The Last Hours of Ancient Sunlight* (New York: Three Rivers Press, 2004), p. 158.

3. Anderson, *Tending the World,* p. 64.

4. John Bellamy Foster, *The Vulnerable Planet: A Short Economic History of the Environment* (New York: Monthly Review Press, 1994).

5. William Roscoe Thayer, *Theodore Roosevelt: An Intimate Biography* (Boston: Houghton Mifflin, 1919), p. 20. http://www.bartleby.com/170.

6. Theodore Roosevelt, *An Autobiography* (New York: Macmillan, 1913), p. 17.

7. Roosevelt, *Autobiography,* p. 430.

8. Roosevelt, *Autobiography,* p. 418.

9. Forest History Society, www.foresthistory.org, http://www.foresthistory.org/research/usfscoll/people/Pinchot/Pinchot.html.

10. Gifford Pinchot, *The Fight for Conservation* (New York: Doubleday, Page, 1910), p. 42.

11. Sierra Club, John Muir Exhibit, www.sierraclub.org, http://www.sierraclub.org/john_muir_exhibit.

12. Michael B. Smith, "The Value of a Tree: Public Debates of John Muir and Gifford Pinchot," *The Historian* 60, no. 4, (June 1998).

13. John Muir, *My First Summer in the Sierra* (Boston: Houghton Mifflin, 1911), p. 20.

14. Roosevelt, *Autobiography,* p. 322.

15. Library of Congress Web site, "The Evolution of the Conservation Movement," http://memory.loc.gov/ammem/amrvhtml/conshome.html.

16. Roosevelt, *Autobiography,* p. 447.

17. Kelly Duane, *Monumental* (Loteria Films, 2005).

18. David Brower, *Let the Mountains Talk, Let the Rivers Run* (New York: Harper-Collins, 1995), p. 9.

19. The Wilderness Society, www.wilderness.org, http://www.wilderness.org/Our Issues/Wilderness/act.cfm.

20. Carolyn Merchant, "Shades of Darkness: Race and Environmental History," *Environmental History* 8, no. 3 (July 2003), http://www.historycooperative.org/journals/eh/8.3/merchant.html.

21. Linda Lear, *Rachel Carson: Witness for Nature* (New York: Holt, 1997), pp. 60–72.

22. Rachel Carson, *Silent Spring* (Boston: Houghton Mifflin, 1962), p. 7.

23. Lear, *Rachel Carson*, pp. 414–30.

24. Ohio History Central, www.ohiohistorycentral.org, http://www.ohiohistorycentral.org/entry.php?rec=1642.

25. Jack Lewis, "The Birth of the EPA," *EPA Journal* (November 1985), http://www.epa.gov/history/topics/epa/15c.htm.

26. Population Connection, www.zpg.org, http://www.zpg.org/media/upload/1219thirtyyears.pdf.

27. Environmental Protection Agency, www.epa.gov, http://www.epa.gov/superfund/about.htm.

28. United Church of Christ, www.ucc.org, http://www.ucc.org/about-us/archives/pdfs/toxwrace87.pdf.

29. Federation for American Immigration Reform, www.fairus.org, www.fairus.org/site/PageServer?pagename=about_about4819.

30. *Special Issue on Environmental Equity, National Law Journal*, September 1992.

31. Known as the father of the environmental justice movement, Robert Bullard is a professor of sociology and the author of twelve books, the first of which, *Dumping in Dixie*, came out in 1990, excerpted here: "Unlike their white counterparts, black communities do not have a long history of dealing with environmental problems. Blacks were involved in civil rights activities during the height of the environmental movement, roughly during the late 1960s and early 1970s. Many social justice activists saw the environmental movement as a smoke screen to divert attention and resources away from the important issue of the day—white racism. On the other hand, the key environmental issues of this period (e.g., wildlife and wilderness preservation, energy and resource conservation, and regulation of industrial polluters) were not high priority items on the civil rights agenda.

"Social justice, political empowerment, equal education, fair employment practices, and open housing were major goals of social justice advocates. It was one thing to talk about "saving trees" and a whole different story when one talked about "saving low-income housing" for the poor. As a course of action, black communities usually sided with those who took an active role on the housing issue. Because eviction and displacement are fairly common in black communities (particularly for inner-city residents), decent and affordable housing became a more salient issue than the traditional environmental issues. Similarly, unemployment and poverty were more pressing social problems for African Americans than any of the issues voiced by middleclass environmentalists.

"In their desperate attempt to improve the economic conditions of their constituents, many civil rights advocates, business leaders, and political officials directed their energies toward bringing jobs to their communities by relaxing enforcement of pollution standards and environmental regulations and sometimes looking the other way when violations were discovered. In many instances, the creation of jobs resulted in health risks to workers and residents of the surrounding communities."

32. Peggy Sheppard founded West Harlem Environmental Action, New York's first environmental justice organization; she was the first female chair of the National Environmental Justice Advisory Council (NEJAC) of the U.S. Environmental Protection Agency. New Orleans resident Dr. Beverly Wright works on behalf of communities in Louisiana's "Cancer Alley"; she heads up the Deep South Center for Environmental Justice at Dillard University.

33. Activist and writer Winona LaDuke (Anishinaabe) has twice been the vice-presidential candidate on the Green Party ticket. Tom B. K. Goldtooth (Dine' and Mdewakanton Dakota) is the national director of the Indigenous Environmental Network, at Bemidji, Minnesota; as he says, "Change will occur when the white men realize their testicles are shrinking—then money will flow like water to the environmental movement." Former chief of the Neetsaii Gwich'in, from Arctic Village in northeastern Alaska, Evon Peter has worked in the United Nations and Arctic Council forum representing indigenous and environmental interests.

34. Richard Moore was one of the organizers of the First National People of Color Environmental Justice Summit in 1991. He helped launch the first successful campaign in New Mexico to clean contaminated groundwater. Today he is he executive director of the Southwest Network for Environmental and Economic Justice. A former migrant farmworker and a garment worker, Alicia Marentes has been an advocate for farmworkers for over thirty years, particularly along the U.S.-Mexico border. In 1997 she was awarded the Letelier-Moffitt National Human Rights Award.

35. Peggy Saika and Pamela Chiang cofounded the Asian Pacific Environmental Network (APEN), an environmental justice organization that builds grassroots leadership in immigrant Asian communities. Vivian Chang is the current executive director of APEN.

36. "On the Road from Environmental Racism to Environmental Justice," *Villanova Environmental Law Journal* 5, no. 2 (1994).

37. Julian Agyeman, *Sustainable Communities and the Challenge of Environmental Justice* (New York: New York University Press, 2005), p. 19.

38. Agyeman, *Sustainable Communities,* p. 36.

39. Interview by Amanda Griscom Little of Majora Carter for www.grist.org, September 28, 2006, http://www.grist.org/news/maindish/2006/09/28/m_carter/index.html.

40. LOHAS Journal, www.lohas.com, http://www.lohas.com/journal/article_archives.html.

CHAPTER 3: ECO-EQUITY

1. President Ronald Reagan, Inaugural Address, January 20, 1981, http://www.reaganlibrary.com/reagan/speeches/first.asp.

2. President Bill Clinton, State of the Union Address 1996, http://clinton4.nara.gov/WH/New/other/sotu.html.

3. Theophilus Eugene "Bull" Connor was the Public Safety Commissioner of Birmingham, Alabama, in the 1960s. A Klansman and staunch advocate of racial

segregation, he appeared on national television spewing anti-integration rhetoric and using attack dogs and fire hoses against (anti-segregation) protestors. A brief video clip of his blather can be viewed at http://www.pbs.org/wgbh/amex/eyesontheprize/story/07_c.html.

4. National Energy Information Center (NEIC), www.eia.doe.gov, http://www.eia.doe.gov/oiaf/1605/ggccebro/chapter1.html.

5. Roy Walmsley, "World Prison Population List, 7th edition," International Centre for Prison Studies, School of Law, King's College London, 2007, http://nicic.org/Library/022140.

CHAPTER 4: THE GREEN NEW DEAL

1. Arthur M. Schlesinger, Jr., *The Coming of the New Deal* (Boston: Houghton Mifflin, 2003), pp. 6–8.

2. Member Survey, Women's Business Enterprise National Council, March 20, 2008, www.wbenc.org, http://www.wbenc.org/PressRoom/News/2008_survey_final.aspx?AspxAutoDetectCookieSupport=1.

3. Step It Up, www.stepitup2007.org; Focus the Nation, www.focusthenation.org; Energy Action Coalition, www.energyactioncoalition.org.

4. The League of Young Voters, www.theleague.com; Hip Hop Caucus, www.hiphopcaucus.org; Environmental Justice and Climate Change Initiative, www.ecc.org; Young People For, www.youngpeoplefor.org.

5. Omar Freilla founded Green Worker Cooperatives in the South Bronx, an incubator for cooperatively run green-collar ventures. In 2007 he won the Jane Jacobs Medal for New Ideas and Activism, awarded by the Rockefeller Foundation.

6. David R. Baker, "Economists Weigh Prop. 87 Arguments," *San Francisco Chronicle*, October 15, 2006, http://www.sfgate.com/cgi-bin/article.cgi?f=/c/a/2006/10/15/BUGMKLOCFH1.DTL.

7. Marc Geller, "California's Proposition 87—What Went Wrong?" www.pluginamerica.org, November 1, 2006, http://www.pluginamerica.org/news-and-press/newsletters/2006-archive/2006-11-01-prop-87-missed-opportunity.html.

8. Grover Norquist on National Public Radio, May 25, 2001, http://www.npr.org/templates/story/story.php?storyId=1123439.

9. Bruce Springsteen on the Vote for Change Tour, October 10, 2004.

CHAPTER 5: THE FUTURE IS NOW

1. U.S. Department of Labor Bureau of Statistics, "Employment Situation Summary," May 2008, http://www.bls.gov/news.release/empsit.nr0.htm.

2. Bracken Hendricks and Jay Inslee, *Apollo's Fire* (Washington, DC: Island Press, 2008).

3. Carla Din, "Finding Opportunity in Crisis," *Yes Magazine* (Fall 2004), http://www.yesmagazine.org/article.asp?ID=1030.

4. Interview with Elsa Barboza, February 2008.

5. Elsa Barboza, "Organizing for Green Industries in Los Angeles," *Race, Poverty and the Environment* 13, no. 1 (Summer 2006), http://www.urbanhabitat.org/node/525.

6. In June 2007, the city council of LA established a City Retrofit Jobs Task Force made up of Apollo Alliance representatives, council members, and employees of various City agencies. The task force is identifying the workforce needs, potential job-training providers, and funding sources.

7. Joanna Lee, Angela Bowden, and Jennifer Ito, *Green Cities, Green Jobs*, May 2007, http://www.greenforall.org/resources/green-cities-green-jobs-by-joanna-lee-angela/download.

8. Lee, Bowden, and Ito, *Green Cities, Green Jobs*.

9. Lee, Bowden, and Ito, *Green Cities, Green Jobs*.

10. Center on Wisconsin Strategy (COWS), http://www.cows.org/collab_projects_detail.asp?id=54.

11. Speech by Winona LaDuke at the Dole Institute, University of Kansas, March 31, 2008.

12. Speech by Winona LaDuke at her acceptance of the nomination as Green Party vice-presidential candidate, July 20, 2000.

13. Indigenous Environmental Network, www.ienearth.org.

14. Honor the Earth, www.honortheearth.org.

15. Native Wind, www.nativewind.org.

16. PRNewswire, "U.S. Steel Permanently Closing Most Fairless Facilities," August 14, 2001.

17. Natalie Kostelni, "Progress Being Made at U.S. Steel Bucks Industrial Site," *Philadelphia News Journal*, January 22, 2007, http://www.bizjournals.com/philadelphia/stories/2007/01/22/story6.html.

18. Steven Greenhouse, "Millions of Jobs of a Different Collar," *New York Times*, March 26, 2008.

19. Karen Breslau, "The Growth in 'Green-Collar' Jobs," *Newsweek*, April 8, 2008, http://www.newsweek.com/related.aspx?subject=Technology.

20. "City Crime Rankings," www.morganquitno.com.

21. City of Richmond, http://www.ci.richmond.ca.us/index.asp?nid=1353.

22. All references to Solar Richmond are drawn from interviews with Michele McGeoy, February 2008.

23. Interview with Angela Greene, February 2008.

24. David R. Baker, "Solar Industry Needs Workers," *San Francisco Chronicle*, May 10, 2008, http://www.sfgate.com/cgi-bin/article.cgi?f=/c/a/2008/05/10/BUGD10JVGP.DTL.

25. Interview with Lyndon Rive, June 2008.

26. SolarCity is also planning to bring solar job opportunities to another depressed part of the San Francisco Bay Area: the Bayview–Hunters Point neighborhood. Bayview has been devastated for years by unemployment, crime, and a dirty coal power plant that was finally shut down in 2006. The City of San Francisco has launched an effort to invigorate the area with clean, green companies, and Solar-City is taking the opportunity to open a training academy there. The company plans to train dozens of new employees every month when the academy opens in late 2008.

27. Lyndon Rive further explains the SolarLease concept: "With SolarLease, the homeowner can pay for the system out of the savings on their electric bill. The

lease customer typically provides a low down payment—say $1,000 or $2,000. After that, the monthly lease payment combined with the new electric bill will typically be less than the old electric bill. We're able to capitalize on commercial tax credits and pass the savings on to the homeowner in the form of lower monthly payments. Most people want to do something positive for the environment, but it needs to make financial sense." Interview with Lyndon Rive, June 2008.

28. Lisa Hymas, "We Built This SolarCity," April 11, 2008, http://www.grist.org/feature/2008/04/11/.

29. Daniel Yee, "Obesity Raising Airline Fuel Costs," November 9, 2004, http://www.livescience.com/health/obesity_airlines_041105.html.

30. Anuradha Mittal, speech at the San Francisco Food Professional Society, Commonwealth Club, San Francisco, October 22, 2004, http://www.oaklandinstitute.org/?q=/node/view/100.

31. Diana Deumling et al., "Eating Up the Earth: How Sustainable Food Systems Shrink Our Ecological Footprint," *Agriculture Footprint Brief* (July 2003).

32. United States Department of Agriculture, Economic Research Service, 2006, http://www.ers.usda.gov/Briefing/FoodSecurity/trends.htm.

33. Interview with Brahm Ahmadi, February 2008.

34. The food system also results in a multitude of health problems as our endocrine, immune, and other systems are besieged by pesticides used on crops and the antibiotics and growth hormones with which (usually healthy) livestock is egregiously sprayed. We're seeing a preponderance of allergies, due in part to the lack of diversity of our food crops, as industrial farms grow mostly just one kind of wheat or corn, for example, out of the many strains that exist or used to exist. See Elizabeth Frazao, "High Costs of Poor Eating Patterns in the United States," *Agriculture Information Bulletin* No. 750 (1999).

35. U.S. Department of Labor, *The National Agricultural Workers Survey,* http://www.doleta.gov/agworker/report/ch1.cfm.

36. Christopher D. Cook, *Diet for a Dead Planet* (New York: New Press, 2004), http://www.dietforadeadplanet.com.

37. Interview with LaDonna Redmond, April 2008.

38. Interview with Orrin Williams, February 2008.

39. Growing Home, www.growinghomeinc.org.

40. The Field Museum, *George Washington Carver Educator Guide* (Chicago: The Field Museum, 2008), p. 18.

41. All references to the People's Grocery and the food crisis in West Oakland as well as quotations are drawn from interviews with Brahm Ahmadi, February 2008.

42. Mittal, speech, October 22, 2004. Mittal clarifies the statistics: "There is not a direct correlation between the annual food expenditures and the income to farmers, as not all dollars spent on food go to farmers. These numbers are based on a research paper by Dennis Tootelian for the California Department of Food and Agriculture's Buy California program."

43. Annie Leonard, "The Story of Stuff," http://www.storyofstuff.com.

44. Justin Berton, "Continent-size Toxic Stew of Plastic Trash Fouling Swath of Pacific Ocean," *San Francisco Chronicle,* October 19, 2007, http://www.sfgate.com/

cgi-bin/article.cgi?f=/c/a/2007/10/19/SS6JS8RH0.DTL&hw=pacific+patch&sn=001&sc=1000.

45. Leonard, "The Story of Stuff."

46. Meanwhile, the companies primarily responsible for handling waste are making a killing. The monopolization of the industry has the few entities in control extorting higher and higher rates for disposal. The nonprofit Institute for Local Self-Reliance calls it "waste imperialism," which "diminishes democratic local ownership and control of valuable discarded materials . . . hampering recycling and waste reduction progress, promoting the interstate transportation of waste, tightening already slim municipal budgets, and sounding the death knell for recycling-based community development and localism in the solid waste sector." See the Institute for Local Self-Reliance, "Waste Imperialism," http://www.ilsr.org/recycling/wasteimperialism.html.

47. One of the smartest approaches to waste reduction is making the companies who produce the stuff responsible for the damages caused by production processes as well as for the fate of the product and its packaging. This principle is called Extended Producer Responsibility (EPR). See the Institute for Local Self-Reliance, "Extended Producer Responsibility," http://www.ilsr.org/recycling/epr/index.html.

48. Jim Motavalli, "Zero Waste," *emagazine* 12, no. 2 (March/April 2001), http://www.emagazine.com/view/?506.

49. Raquel Pinderhughes, *Green Collar Jobs,* The City of Berkeley Office of Energy and Sustainable Development, 2007, http://www.bss.sfsu.edu/raquelrp/documents/v13FullReport.pdf.

50. Information about Chicago's computer recycling from an interview with Patrick Brown of OAI/Greencorps Chicago, February 2008.

51. Pinderhughes, *Green Collar Jobs.*

52. Environmental Protection Agency, "Characterization of Building-Related Construction and Demolition Debris in the United States," www.epa.gov/epaoswer/hazwaste/sqg/c&d-rpt.pdf.

53. Doug Brown, "Search and Rescue," *Washington Post,* June 19, 2003.

54. All references to Green Worker Cooperatives and ReBuilders Source from interview with Omar Freilla, May 2008.

55. WaterAid, "Statistics," http://www.wateraid.org/international/what_we_do/statistics/default.asp.

56. Storm-water runoff is the result of human intervention in the hydrologic cycle in the course of urban and suburban development. Impervious surfaces, soil compacting, and removal of vegetation alter water's movement through the environment by reducing interception, evapo-transpiration, and infiltration. Storm water is a major source of pollution: in the United States it's the largest pollution source for ocean shorelines, second largest source for estuaries and Great Lakes shorelines, third largest source for lakes, and fourth largest source for rivers. Traditional storm-water management practices focus on the collection and rapid removal of rainwater and snowmelt away from the point of impact through a system of underground pipes and storm sewers. This approach generates vast quantities of polluted runoff, disrupts the natural hydrologic cycle, and adds to the contami-

nation and scouring of streams and rivers. See the Water Environment Research Foundation, http://www.werf.org/livablecommunities/tool_bmp101.htm.

57. Interview with Majora Carter, April 2008.

58. Interview with James Wells, February 2008.

59. Maude Barlow and Tony Clarke, *Blue Gold: The Fight to Stop the Corporate Theft of the World's Water* (London: Earthscan, 2002), p. 233.

60. Barlow and Clarke, *Blue Gold*, p. 235. Another major solution to the global water crisis involves shifting crops to match the availability of local water supplies, which means no longer growing water-intensive crops in arid climates.

61. All references to TreePeople and its LA watershed projects from interviews with Andy Lipkis, May and June 2008.

62. Surface Transportation Policy Project, *Factsheet on Transportation and Climate Change*, http://www.transact.org/library/factsheets/climate.asp.

63. Interview with Sam Zimmerman-Bergman, May 2008.

64. Surface Transportation Policy Project, *Factsheet on Transportation and Social Equity*, http://www.transact.org/library/factsheets/equity.asp.

65. American Public Transportation Association, *Public Transportation Industry Overview*, http://www.apta.com/media/facts.cfm.

66. Surface Transportation Policy Project, *Factsheet on Transportation and Jobs*, http://www.transact.org/library/factsheets/jobs.asp. In fact, according to a study by Cambridge Systematics, although new highway construction does lead to an increase in employment, these jobs are mostly for nonlocal workers: road engineers and other specialists who come in to an area for a specific job and then leave when it has been completed. On the other hand, transit investments create a wealth of employment opportunities in the short and the long run. Transit system construction leads to an impressive level of short-term job creation, and once the systems are finished, a long-term source of high-quality jobs.

67. J. Mijin Cha, "Public Transit Needed in California and Across Nation: California Progress Report," Transportation Equity Network, February 26, 2008, http://transportationequity.org/index.php?option=com_content&task=view&id=62&Itemid=45.

68. Darrell Clarke, "Curitiba's 'Bus Rapid Transit'—How Applicable to Los Angeles and Other U.S. Cities?" *Light Rail Now* (March 28, 2005), http://www.lightrailnow.org/facts/fa_00013.htm.

69. "Solis' Green Jobs Act Signed into Law as Part of Historic Energy Reform Bill," December 19, 2007, http://solis.house.gov/list/press/ca32_solis/wida6/greenjobslaw.shtml.

CHAPTER 6: THE GOVERNMENT QUESTION

1. Civilian Conservation Corps Legacy, http://ccclegacy.org.

2. Stan Cohen, *The Tree Army: A Pictorial History of the Civilian Conservation Corps, 1933–1942* (Missoula, MT: Pictorial Histories, 1960).

3. John C. Paige, "The Civilian Conservation Corps and the National Park Service, 1933–1942: An Administrative History," National Park Service, 1985, http://www.nps.gov/history/history/online_books/ccc/index.htm.

4. Franklin Delano Roosevelt, Greetings to the CCC speech, July 1933, http://www
.parks.ca.gov/?page_id=24917.

5. James F. Justin Civilian Conservation Corps Museum Biographies, http://members
.aol.com/famjustin/cccbio.html.

6. John F. Kennedy, space-program address at Rice University, September 12, 1962,
www.astrosociology.com/Library/PDF/JFK%201962%20Rice%20University%20
Speech%20Transcript.pdf.

7. Bracken Hendricks and Jay Inslee, *Apollo's Fire* (Washington, DC: Island Press,
2008), p. xviii. The slogan of the Apollo Alliance is "Three million new jobs. Freedom from foreign oil." More at http://www.apolloalliance.org/.

8. The CEC would mobilize millions of Americans of all ages, particularly the four
to six million youth between the ages of sixteen and twenty-four who are neither
in school nor able to land meaningful jobs. Working in collaboration with business, labor, schools, and other components of the green economy, the CEC would
build on existing service and conservation corps and green-jobs training programs. Regional and community green-jobs training programs need to be vastly
expanded to meet the need for skilled workers. An investment of $5 billion a year
would support matching grants to states and cities for green workforce development, with a focus on job preparation, matching, and retention efforts for the
unemployed or poor. One billion dollars a year would mobilize twice the current
number of AmeriCorps members in a dedicated CEC national service corps.

   The CEC would have recruiting centers in cities across the country, connecting people to work and service opportunities generated by the implementation
of national, state, and local climate-change policies. CEC could retrofit our
aging buildings; green America's public elementary, middle, and high schools;
expand green space in cities, transforming blacktops into parks or urban farms;
assist in food composting and recycling; and build bike trails and organize car-
pools. Corps members would gain concrete experience, hard and soft skills, and
be ready to continue their education, enter preapprenticeship programs, or start
their own businesses after leaving the corps.

9. Susan Tucker, director of the After Prison Initiative at the Open Society Institute,
is a leading voice on connecting formerly incarcerated people with green-job
opportunities. Susan's goal is to scale and institutionalize the CJC as a major
federally funded program and to establish a local CJC in every high-incarceration
neighborhood.

   It may be hard to understand the appeal of such a program, unless you understand the devastating impact on communities of color of the mushrooming
incarceration industry. According to the Urban Strategies Initiative, over the
past thirty years the United States has become the number-one spender on
incarceration in the world, locking up more people and a bigger percentage of
its population than Russia or China. Most of the jailed are people of color, the
majority for nonviolent offenses—essentially crimes of economic desperation,
addiction, and mental illness. Outside of prison there are now 12 million formerly incarcerated people, whose number grows by half a million a year, while
more than 40 percent of formerly incarcerated people are returned to prison
within three years of release.

A training program with the wraparound services necessary to ensure success for the truly disadvantaged might cost as much as $20,000 per person. But compare that to the $40,000–$100,000 we spend per person in California's prisons—on a system that utterly fails in its "correctional" purpose. According to the Urban Strategies Initiative, more than two-thirds of all adult prison parolees are rearrested within three years of release; an astounding 91 percent of juveniles reoffend. (In both cases, many of those who return to prison are arrested on technical violations or because of legal barriers to their housing, employment, education, and treatment due to their conviction.) The success rate of corps programs that serve comparable populations suggest that a $20,000 investment is a bargain. The mass scale payoffs in reduced costs for prison and social welfare programs plus the gains in peaceful streets and redeemed lives would be simply incalculable.

10. Though variations of the idea are controversial on the left and the right, a cap on carbon is generally thought to be the most effective macro-tool to help society cut greenhouse-gas emissions. A federally instituted cap would set a limit on the total amount of carbon that could be emitted in the United States. Most proposals then call for the government to issue permits that would allow a certain amount of emissions each year. And every year, the number of permits—and therefore the amount of pollution permitted—would decrease. This process would bring down our emissions steadily but gradually, since no one believes a drastic reduction is practical.

And yet those who advocate for carbon trading argue that a "cap-and-trade" system would be a highly efficient and flexible way to cut emissions. In their view, the problem is a simple market malfunction: dumping carbon into the atmosphere has always been free for any individual or firm—and yet dumping all that carbon could ultimately cost us the entire planet. Advocates argue that once businesses are forced to pay the "true cost" of carbon (by buying and trading permits), they will have the necessary financial incentive to seek out low-carbon solutions. Then the magic of the market will kick in and start moving us quickly to safer ground.

As for me, I do not believe that the climate crisis is simply a market failure, nor do I believe that market solutions alone will save us. The climate crisis is a symptom of a much deeper sickness in our souls and our society. I find it distasteful to think of anyone "buying" the right to pollute the planet. And yet I can see few other politically viable, practically workable options. For instance, there is no political will, at this stage, to treat carbon like cyanide and just ban it altogether. And we couldn't do so without plunging ourselves into the Dark Ages anyway.

So in my opinion, as we grasp for a solution, three values are essential. One, we must bring down carbon emissions as fast as possible. Two, the big polluters must start paying for their actions somehow (through carbon taxes, carbon tariffs, buying permits, or even fines). And three, the billions generated by pricing carbon must go toward helping the people and the planet.

Therefore, I propose that we proceed by affirming a new concept: "cap and collect and invest." To ensure equal protection and equal opportunities for everyone, we need the government to *cap* carbon (set a firm limit), *collect* money from big emitters (either by a carbon tax, by selling the limited number of pollution permits), and *invest* the funds (spend them on projects that will further boost the clean, green economy, while protecting people from the worst economic consequences of the changeover).

Neither the carbon-tax ideas nor the permit-trading schemes are perfect. But as long as the United States establishes a firm cap, we could experiment with any number of creative ways to "collect," or make the big polluters pay. But what we don't need, under any circumstances, is for the government to allow polluters to purchase offsets instead of reducing emissions. Offsets are certificates sold by private companies that claim to invest the cost of each certificate in activities that remove carbon from the atmosphere, such as planting trees. Everyone can be encouraged to purchase offsets, but they are no substitute for direct reductions in emissions.

And what we really don't need is for the federal government to institute a "cap and giveaway" policy, in which permits are "grandfathered," that is, given away for free. Although such a policy does still result in a lowering of emissions (providing offsets are not allowed to be substituted for emissions), this proposal rewards longtime polluters, rather than making them take responsibility for their disastrous impacts on public health and the environment. It leaves a big pile of money—in the tens or hundreds of billions of dollars, by various estimates—as windfall profits for major fossil-fuels users.

As we face this global ecological crisis, our largest industries should stop crying and begging for freebie pollution permits—especially our energy companies. We honor their contributions to this nation; they have kept the lights on for us all this time. But in an age of windfall profits for them and worldwide peril for every living species, the time has come for them to let go of their billion-dollar subsidies. The time has come for them to start paying this country back for all the taxpayer support we have given them for decades. They can best begin to do that by paying full price for each and every pollution permit, as an act of patriotic duty.

We should also pass measures to protect U.S. industries, such as a carbon border fee (or carbon tariff), which would impose fees on goods imported from countries where carbon still costs less. And we should also protect vulnerable people as well. The truly needy should get earned income tax credit or direct cash assistance to offset increasing energy and food prices in a carbon-constrained economy.

Of course, everyone will feel the pinch at first. To ease the pain, the government could take part of the money collected from carbon proceeds and send a check to every U.S. citizen. Not only would those checks help everyone defray some of the costs of increased energy prices and prices of basic consumer goods, they also would give every U.S. citizen a financial stake in the transition to a cleaner economy. But I do not believe we should give away *all* of the money. The bulk of the money collected from pricing carbon should go to pay for robust programs to build green infrastructure and create green-collar jobs across the country.

An exhaustive description of cap-and-trade is beyond the scope of this chapter, but can be found in CAP's seminal report, *Capturing the Energy Opportunity,* http://www.americanprogress.org/issues/2007/11/pdf/energy_chapter.pdf.

11. Congressional Budget Office, *Trade-Offs in Allocating Allowances for CO_2 Emissions,* August 2007, http://www.cbo.gov/ftpdocs/80xx/doc8027/04-25-Cap_Trade.pdf.

12. PJM Interconnection, *Bringing the Smart Grid Idea Home,* www.energyfuture coalition.org/pubs/PJMsmartgrid.pdf.

13. John M. Urbanchuk, for the Renewable Fuels Association, *Contribution of the Ethanol Industry to the Economy of the United States,* February 2008, http://www.ethanolrfa.org/objects/documents/1537/2007_ethanol_economic_contribution.pdf

14. Interview with Ross Gelbspan, May 2008. See http://www.heatisonline.org for more information.

15. NATO Secretary General Jaap de Hoop Scheffer, speech at the Security and Defense Agenda, June 3, 2008, http://www.nato.int/docu/speech/2008/s080603a.html.

16. The Greenhouse Development Rights (GDR) framework codifies the right to development as a "development threshold"—a level of welfare below which people are not expected to share the costs of the climate transition. This threshold is emphatically not an "extreme poverty" line, which is typically defined to be so low ($1 or $2 a day) as to be more properly called a "destitution line." Rather, it is set to be higher than the "global poverty line" and to reflect a level of welfare that is beyond basic needs, but well short of today's levels of "affluent" consumption.

People below this threshold are taken as having development as their proper priority. As they struggle for better lives, they are not asked to help keep society as a whole within its now sharply limited global carbon budget. In any event, they have little responsibility for the climate problem and relatively little capacity to work on solving it. People above the threshold, on the other hand, are taken as having realized their right to development and as bearing the responsibility to preserve that right for others. They must gradually assume a greater fraction of the costs of curbing the emissions associated with their own consumption as well as the costs of ensuring that those who rise above the threshold are able to do so along sustainable, low-emission paths. And these obligations, moreover, are taken to belong to all those above the development threshold, whether they happen to live in the North or in the South. See http://www.ecoequity.org/GDRs/.

17. United Nations, Department of Economic and Social Affairs, Population Division, *World Population Prospect: The 2006 Revision,* http://www.un.org/esa/population/publications/wpp2006/wpp2006.htm.

18. Interview with Sadhu Johnston, February 2008.

19. Interviews with Patrick Brown, OAI/Greencorps, and Aaron Durnbaugh, Deputy Commissioner, Chicago Department of Environment, February 2008.

20. Interview with Jumaani Bates, March 2008.

21. COWS is working in partnership with the City of Milwaukee, the local area utility, and community leaders to ensure Me2's success. Me2 will mostly use private

capital in doing this retrofit work. It has already attracted banks and other private creditors to the effort because it pools expected savings that would otherwise be too dispersed to capture. To minimize risk for the lender, it puts the repayment costs of its capital, and the retrofit work done with it, on the customer utility bills, with standard penalties for nonpayment. It also aggregates individual customers for energy services into a single borrowing pool. To attract wide participation, it selects for efficiency measures that will more than pay for themselves and the cost of capital over a reasonable (say, ten-year) period. That means that, even though customers will be repaying the cost of the retrofits, their utility bills will go down immediately.

Here's how Me2 will work. The utility agrees to administer repayment to Me2 as part of its billing services. A bank agrees to loan or open a credit line for capital at a certain rate. Me2 recruits customers willing to pay for approved work. The bank loans Me2 funds for the expected cost of that work. Me2 sends auditors and contractors to the customer's property, gets agreement from the customer on the scope of work, and has work done and verified. The utility bill drops; customer repayment of Me2 costs begins. The utility forwards that repayment to Me2, which repays the bank.

Me2 guarantees the work for customers, with standards for those who do the work. It assures customers of an immediate drop in their energy costs, even with payment to it, by carefully screening the retrofit measures it will pay for.

Crucially, the payment for Me2 services "follow the meter" or property, not person. That means Milwaukee renters and owners would gain from the program without incurring the obligation to stay in their current property. Whatever they haven't repaid for services simply gets passed (with notice of course) to the next renter or owner. For more details see http://cows.org/collab_projects_detail .asp?id=54.

22. Population Reference Bureau, "World Population Highlights 2007," http://www .prb.org/Articles/2007/623Urbanization.aspx.

23. Greg LeRoy, "Green Strings," www.grist.org, May 1, 2008, http://gristmill.grist .org/story/2008/4/30/113724/493.

CHAPTER 7: BUOYANCY AND HOPE

1. Niccolò Machiavelli, *Il Principe (The Prince)* (Florence: Antonio Blado d'Asola, 1532), p. 12.

2. Lester R. Brown, *Plan B 3.0* (New York: Norton, 2008).

3. Winston S. Churchill, *The Second World War* (London: Cassell, 1948–54).

4. Winston Churchill, addressing the House of Commons, May 13, 1940, http:// www.fordham.edu/halsall/mod/churchill-blood.html.

Acknowledgments

First of all, grateful appreciation goes to my parents, Loretta Jean Kirkendoll Jones and the late Willie Anthony Jones, and my grandparents, the late Bishop Chester Arthur Kirkendoll and Alice Elizabeth Singleton Kirkendoll. They gave me the best foundation that any black kid who grew up on the edge of a small Tennessee town could ever hope for.

I honor and acknowledge my life partner, Jana Carter, and our beautiful sons, Cabral and baby Mattai. They have made untold sacrifices so that I could create this book—and do the work that underlies it.

I salute my entire family: the Kirkendolls, the Carters, and the legendary Smith-Jones-Glover clan of Memphis, Tennessee. My twin sister, Angela Thracheryl Jones, and her sons, DeAubrey Jerome and Brandon Demetrious Weekly, are never far from my thoughts.

I also thank Diana Frappier, my steadfast friend, with whom I cofounded the Ella Baker Center for Human Rights in 1996. And I acknowledge my loving godparents, Dorothy Zellner, Constancia "Dinky" Romilly, and Terry Weber.

I thank the many volunteers, supporters, coworkers, cofounders, and board members with whom I have had the pleasure of launching and helping to lead three social change organizations: the Ella Baker Center, Color of Change, and Green For All. We are just getting started.

As a lifelong activist, I have been blessed with scores of friends, allies, and comrades. It would be impossible to name them all here, but any decent list would include Monica Elizabeth Peek, Theda Sevier Hunt, KaCarole Higgins, Karen Streeter, Emilee Whitehurst, Cindy Weisner, Raquel Laviña, Rahdi Taylor, Judy Appel, Craig Harshaw, Lisa Daugaard, Deborah James, Marianne Manilov, Tony Coleman, Priya Haji, Gillian Caldwell, Kolmilata Majumdar, Aya De Leon, Kwame Anku, Michelle Loren Alexander, Bernadette Armand, Alli Chagi Starr, Leda Dederich, Shalini Kantayya, Gita Drury, Billy "Upski" Wimsatt, James Rucker, Tesa Silvestre, Zen DeBruke, Rachel Bagby, Nina Utne, Vivian Chang, Juliet Ellis, Amaha Kassa, Taj James, Gihan Perera, Baye Adolfo-Wilson, Rev. Lennox Yearwood, John Hope Bryant, Bracken Hendricks, Peter Teague, Michelle DePasse, Diane Ives, Kalia Lydgate, Ariane Conrad, Valerie Aubel, Sarah Shanley, Monet Zulpo-Dane, Lea Endres, Majora Carter, and Julia Butterfly Hill.

I am the product of a first rate education, thanks in part to Ollye Curry (Alexander Elementary), Helen Mahafy (Jackson Central-Merry High School), E. Jerold Ogg (University of Tennessee, Martin), and Stephen Wizner (Yale Law School). I have also enjoyed the support of world-class mentors and guides. They include: Max Elbaum, Linda Burnham, Bob Wing, Elizabeth "Betita" Martinez, Sharon Martinas, Belvie Rooks, Kathleen Cleaver, Kerry Kennedy, Arianna Huffington, Samsara Becknell, Catherine Sneed, Fred and Ina Pockrass, Jim Sheehan, David Friedman, George Zimmer, Kevin Danaher, Jerome Ringo, Joel Makower, Michael Kieschnick, Jodie Evans, Lynne Twist, John Robbins, Paul Hawken,

George Lakof, Mal Warwick, Josh Mailman, Arnold Perkins, Joel Rogers, and Robert Gass. I especially honor Eva Jefferson Paterson for giving me my first job as a lawyer and supporting all my dreams ever since.

The insights in this book arise as a part of a long-running conversation within a growing community of dedicated thinkers and activists. I cannot claim to be the sole author or lonely originator of all the ideas presented here. For any valuable contribution readers find herein, I must share the credit with my colleagues at organizations like the Apollo Alliance, 1Sky, Bioneers, the Ella Baker Center, the Blue Green Alliance, the Social Venture Network, Center for American Progress, Green Festivals, BALLE/Business Alliance for Local Living Economies, Beatitudes Society, Awakening the Dreamer, the Institute of Noetic Sciences, the Tipping Point Network, Rockwood Leadership, Turning the Tide Coalition, the Alliance for Climate Protection, and all of Green For All's partner organizations. As for the shortcomings in conception or presentation, those are mine alone.

Ariane Conrad has been my tireless copilot in creating this book. She is especially appreciative of Peter Barnes and the Mesa Refuge, the Green For All staff, the San Francisco Writers Grotto, David Levy and Mayacamas Ranch, Café Flore, especially Matt—and all her patient Loved Ones. I also thank them.

Plus I deeply appreciate and thank my agent, Patti Breitman, as well as Gideon Weil and his wonderful colleagues at HarperOne.

In closing, I would like to note that my father would have enjoyed reading this book—and grilling me on every point. Unfortunately, he died on March 9, 2008—before the writing was complete. With a grieving heart, I dedicate this book to his memory. Rest in peace. Your spirit and your cause live on.

VAN JONES is the founder and president of Green For All. An internationally acclaimed and award-winning human rights and environmental leader, Jones is a Senior Fellow at the Center for American Progress. He co-founded the Ella Baker Center for Human Rights in 1996, and in 2005 helped found ColorOfChange.org, an online advocacy organization. In addition, Jones is a board member of 1Sky, the Apollo Alliance, and a fellow with the Institute of Noetic Sciences. A Yale Law graduate, Van Jones lives in Oakland with his wife and two sons. Visit the author online at www.vanjones.net and www.greenforall.org.

ARIANE CONRAD is a San Francisco-based writer, editor, and activist. She previously edited a collection of Van's writings entitled *The Future's Getting Restless* (2008). Visit her at www.arianeconrad.com.